The Cambridge Manuals of Science and
Literature

AERIAL LOCOMOTION

Farman biplane in flight. (Topical Press Agency.)

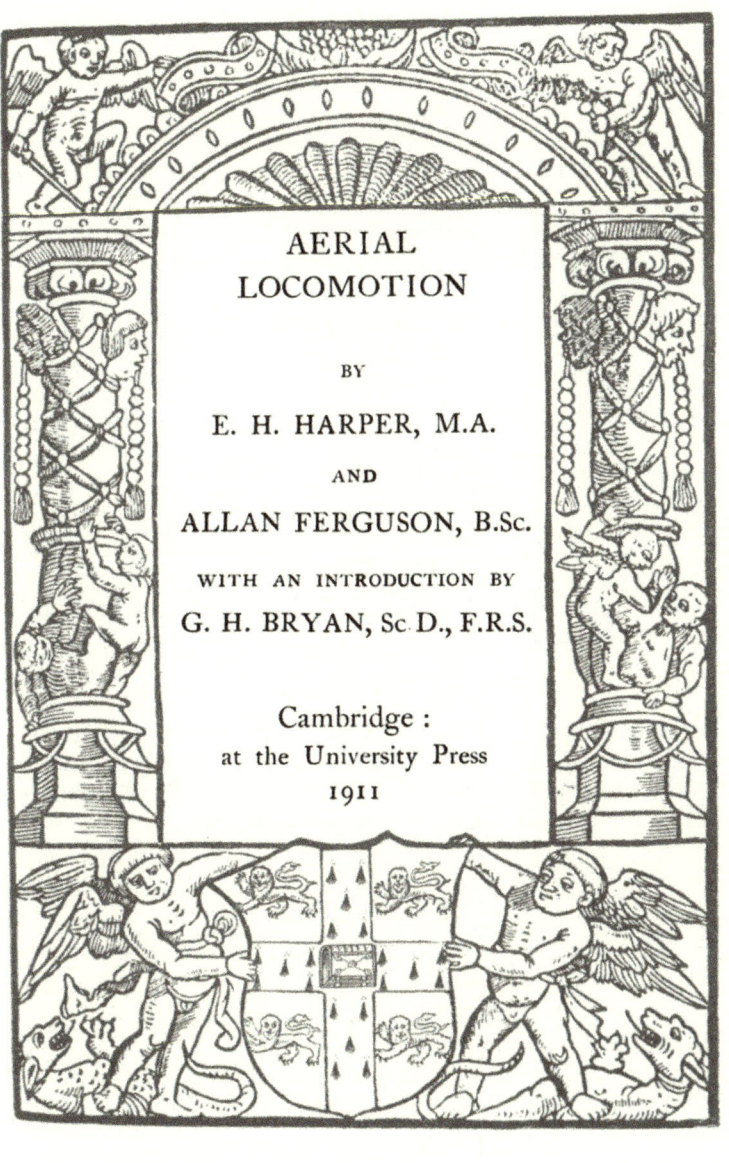

AERIAL LOCOMOTION

BY

E. H. HARPER, M.A.

AND

ALLAN FERGUSON, B.Sc.

WITH AN INTRODUCTION BY

G. H. BRYAN, Sc.D., F.R.S.

Cambridge :
at the University Press
1911

CAMBRIDGE UNIVERSITY PRESS
Cambridge, New York, Melbourne, Madrid, Cape Town,
Singapore, São Paulo, Delhi, Tokyo, Mexico City

Cambridge University Press
The Edinburgh Building, Cambridge CB2 8RU, UK

Published in the United States of America by Cambridge University Press, New York

www.cambridge.org
Information on this title: www.cambridge.org/9781107605923

© Cambridge University Press 1911

First published 1911
First paperback edition 2011

A catalogue record for this publication is available from the British library

ISBN 978-1-107-60592-3 Paperback

*With the exception of the coat of arms at
the foot, the design on the title page is a
reproduction of one used by the earliest known
Cambridge printer, John Siberch, 1521*

PREFACE

THIS little volume makes no claim to originality, its purpose being to present, in a manner as simple as is consistent with scientific accuracy, a connected statement of the principles underlying aerial locomotion ; and for this reason, whilst a certain amount of technical and scientific phraseology has been employed, part of the introductory chapter has been devoted to an elucidation of those mechanical ideas and terms, a correct comprehension of which is absolutely necessary to the reader who desires to follow intelligently the development of modern aeronautics. It will be noticed that some of the chapters 'overlap' to a slight extent—a result which it has not been thought worth while to avoid, as a second presentation of facts which may be somewhat unfamiliar will not be without its value.

It becomes more and more difficult, as the number of manuals and treatises on aeronautics continually increases, to find a title which possesses the merit of freshness. So far as we are aware, the title 'Aerial Locomotion' has only previously be used in a paper by F. H. Wenham, published in 1866.

The frontispiece to the book, and Figs. 13, 14, 20, and 21, have been supplied by the Topical Press Agency; Fig. 22 is from an old print, and Fig. 26 has been taken from Langley's *Researches and Experiments in Aerial Navigation.* Our thanks are tendered to Messrs Whittaker & Co. for permission to reproduce several illustrations from the English translation of Moedebeck's *Pocket-book of Aeronautics.*

E. H. H.
A. F.

University College of North Wales,
  Bangor.
    April, 1911.

CONTENTS

INTRODUCTION

By Professor G. H. BRYAN, Sc.D., F.R.S.

THE accounts of aviation feats contained in news-
papers do not afford the average reader much
information regarding the construction and working
of the machines on which these records have been
performed. One day we may hear of an aviator
winning a prize of £1000 for a flight from Land's End
to John o' Groats, another day we are told that three
Italian aviators have been killed. In some cases
record flights are said to have been made on monoplanes, in others on biplanes, and the question is thus
often asked, which is better, the monoplane or the
biplane? Frequently advantages are claimed for one
or other of these types, which do not really depend
at all on whether the machine is a monoplane or a
biplane, but which result merely from the fact that
some particular machine of one or other type possesses structural peculiarities which other machines
do not possess. Such peculiarities may depend, for
example, on the position of the centre of gravity or of
the propeller, and not on whether the machine has
one lifting surface or two superposed surfaces.

It is the object of Messrs Harper and Ferguson's
book to give an intelligent general account of the
construction and use of aeroplanes and of the way in

which they are adapted to the task they have to perform. A further useful purpose which the book should serve is that of showing the important part which has been played by indirect methods of investigation in rendering aviation possible. From the earliest days of history there has never been any lack of men who would only be too glad to build flying machines and risk their lives by trying to fly them, but until quite recently the means available for accomplishing this object were inadequate. The change was brought about partly by a scientific study of the conditions of the problem, based in the first instance on determinations of the pressure of air on moving planes. This is a subject which had attracted the attention of mathematicians long before aviation became possible, and two alternative theories had been proposed, one due to Newton and the other to Helmholtz and Kirchhoff. The next stage, the experimental one, was mainly developed by the late Dr S. P. Langley who showed that the conditions requisite for aviation were more favourable than had been previously supposed. In this connection, however, the same old story repeated itself, which has so often retarded progress in discoveries and inventions, namely prejudice against the scientific expert. The grant available for Dr Langley's experiments was withdrawn just at the time when success was imminent. Had this not been the case it is probable that the development of modern aviation might have proceeded along safer lines than has actually been the case. For it is certain that the models with which Dr Langley experimented possessed better inherent stability than the majority of existing aeroplanes, otherwise they would not have performed long flights without overturning in some way or other.

The advantages of theoretical and indirect methods were further illustrated in the case of aeroplane motors. It was largely the development and improvement of motors for use on roads which led to the evolution of a motor sufficiently light and powerful for the purposes of aviation. But it cannot be denied that the construction of internal combustion engines has been greatly facilitated by the development of the science of thermodynamics which long preceded their introduction.

Although there is still a good deal of hostility on the part of some practical men towards workers in pure science, there are nevertheless indications that a better state of feeling is beginning to arise, and that those who are concerned with the construction and use of aeroplanes are beginning to show greater willingness to pay attention to investigations of a theoretical character which may be likely to suggest directions for further improvement. The present book will I hope show the general reader that the problem of flight is a many-sided one and that it is only by studying this problem from every possible point of view that locomotion through the air can be brought to the same final stage of development that has been attained in the case of locomotion on land or water.

CHAPTER I

GENERAL PRINCIPLES

THE conquest of the elements around him is, and ever has been, one of the objects for which man is continually striving. He early learned to support himself in water and to move rapidly over its surface, but the emulation of the birds, though dreamed of since the days of Dædalus, has only of recent years become an accomplished fact. And now, when the subject of artificial flight is so much, literally and metaphorically, 'in the air,' when each day's newspaper has its account of some new record made, or old record broken, an elementary account of the principles underlying artificial flight, and the main types of machine in common use, may prove useful to those readers who are not content to take facts as they stand, but wish to know something both of the 'why' and the 'how' of those facts.

It is however, impossible, to demonstrate clearly these principles without some knowledge of the fundamentals of mechanics, and as an exposition of the points which chiefly concern us will, in the future, save much circumlocution, we give them here by way of preamble.

We shall constantly use the term 'force,' and since in aeronautics, as in all scientific work, it is of primary importance to avoid looseness of terminology, we follow Newton's example, and define the term 'force' to mean 'that which alters, or tends to alter, a body's state of rest, or uniform motion in a straight line.'

If, then, any body under consideration be either in a state of rest, or moving with a constant speed, we infer that no forces are acting on it. If, on the other hand, its speed is varying *either in magnitude or in direction*, we infer that some force is acting on the body and causing this variation. The importance of the phrase which we have just emphasised will be made apparent in the sequel.

Now, in order to specify a force completely, we require to know both its magnitude and its direction. A force equal to the weight of 10 lbs. acting in a northerly direction is very different from a force of 10 lbs. weight acting in an easterly direction, for the one will urge the body on which it acts towards the North, the other towards the East. Bearing this in mind, we see that we can graphically depict any given force by means of a straight line, the length of the line, on some appropriate scale, representing the

magnitude of the force, and its direction the direction
of the force.   Thus in the diagram, the line *AB* may
be taken to represent a force of 10 lbs. weight acting
to the North, the line *AC* a force of 8 lbs. weight
acting to the North-East, and the line *AD* a force of

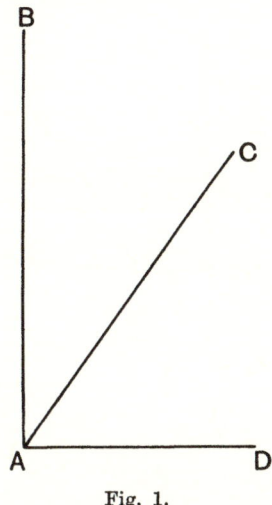

Fig. 1.

5 lbs. weight acting to the East.   Now, if only one
force be acting on a body, that body will move with
ever increasing speed in the direction in which the
force is acting.   But supposing that at some particular
point of a body we have *two* forces applied ; how will

1—2

the body move under these circumstances? The
answer is given by a very simple construction, which
the reader is asked to bear carefully in mind, as we
shall continually make use of it. Suppose *A* to be
the point in the body at which the two forces are
applied, and let the magnitude and direction of action

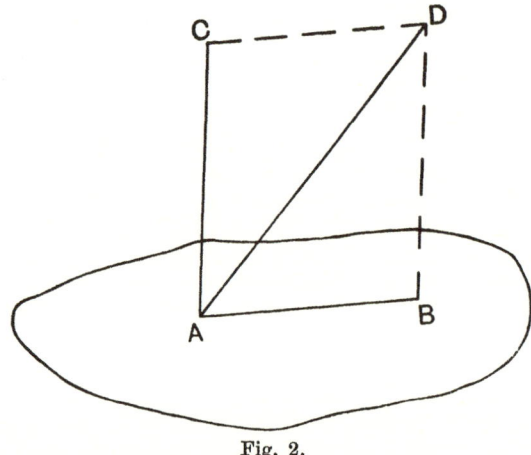

Fig. 2.

of one force be represented by the line *AB*, and of
the other by the line *AC*.

Draw the dotted line *CD* parallel to *AB* and *BD*
parallel to *AC*, so that these two lines meet at *D*.
Draw the straight line *AD*. Then the body will
move in the direction *AD*, and its motion will be

such as if a *single* force of magnitude AD were acting on it. (AD will be drawn on the same scale as AB and AC.) And *vice versa*, any *single* force such as AD acting on a body can be split up into *two* forces such as AB and AC, and the effect of each of these forces studied separately. In scientific terminology, AD is called the *resultant* of the forces AB and AC, and the forces AB and AC are called *components* of the force AD.

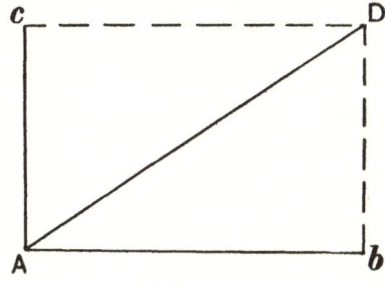

Fig. 3.

It is pretty clear that a force such as AD can be divided into two 'components' in any number of ways, for any number of parallelograms can be drawn having AD as diagonal, but when the parallelogram becomes a rectangle, as in Fig. 3, then Ab and Ac are called, *par excellence*, *the* components of the force AD, and whenever we use the phrase '*the* components' it will be with this signification.

Now let us study in some little detail the motion of a body which is acted on by several forces. Suppose, for example, we have a body acted on by two vertical forces P and Q, one acting upwards, the other downwards, the points at which these two forces act being in a vertical line. Then the actual force—

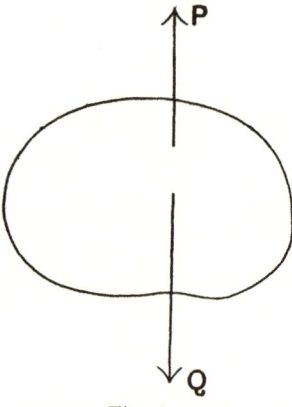

Fig. 4.

the *resultant* force—on the body will be a force equal in magnitude to the difference between P and Q, and the body will move upwards or downwards according as P is greater or less than Q. If P be exactly equal to Q, then, of course, the body will be in equilibrium.

It is important to notice, also, that unless the

points at which the two forces act are in a vertical straight line, the conclusions which we have just reached will no longer be valid. If the two forces, which we will suppose to be equal in magnitude, act as shown in Fig. 5, then, so far from the body being in equilibrium, it will turn round in the direction

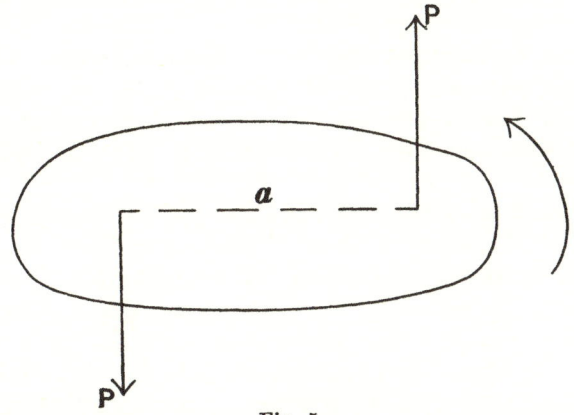

Fig. 5.

indicated by the arrow, and the correct magnitude of this turning tendency is measured by the product of one of the forces into a, the perpendicular distance between the two forces. In fact the two forces together constitute a turning system, which is known as a 'couple.'

At this juncture it is advisable to say a few

words about the famous 'Third law of motion' first
enunciated by Newton. It may briefly be expressed
thus—'To every action there is an equal and opposite
reaction.' For example, if we press our hands upon
a table, the table exerts an equal and opposite force
on our hands. If a horse in pulling a cart exerts a
definite force upon it, the cart exerts an equal and
opposite force upon the horse. This equality of
action and reaction has been made an occasion for
a well worn and somewhat absurd puzzle-question—
'Why, if these two forces are equal, do the horse and
cart ever get into motion at all?' The answer is
obvious. In dealing with the motion of any body,
we must ask ourselves what are the forces acting on
that body, *and on that body only*. The sum of
these forces when added, or compounded, in the
manner shown above will, in general, be a definite
force acting in a definite direction, and the body will
move in that direction with ever-increasing velocity.
Now, returning to our problem of the horse and cart,
the forward pull of the horse is exerted *on the cart*,
whilst the backward reaction of the cart is exerted
on the horse, and has nothing to do with the motion
of the cart itself. It is, indeed, a fact which is often
lost sight of, that all forces consist of *stresses* between
two material bodies, and the Newtonian mode of
estimating forces tends to fix attention on one aspect
only of the stress.

And now, leaving these somewhat tedious, but very necessary, introductory principles, let us see how they may be applied to the elucidation of aeronautical problems.

Before we can propel ourselves through the air, we must first be able to rise into the air, and as is well known this latter problem is solved by the use of two types of machine—the heavier-than-air machine, and the lighter-than-air machine. The kite is a representative of the former class, the balloon of the latter.

Let us deal with the latter type first, and seek an answer to the question 'Why does a balloon rise from the ground?'

Long before a balloon ever did rise—2000 years ago—the answer was supplied by Archimedes, who showed that when a solid is completely immersed in any fluid, it experiences an upward thrust which is equal to the weight of fluid displaced.

Think, then, of a solid substance immersed in a fluid such as water. Two forces act on it—its weight Q in a vertically downward direction (see Fig. 6), and the upward force or thrust P, which by the principle of Archimedes is equal to the weight of fluid displaced.

Now if the solid, volume for volume, be heavier than the fluid, Q will be greater than P; the resultant force will therefore be downwards, and the body will

sink. If the solid be specifically lighter than the
fluid, *P* will be greater than *Q* and the body will rise
to the surface, while, if the solid and fluid have the
same specific gravities, the solid will rest indifferently
in any part of the fluid.

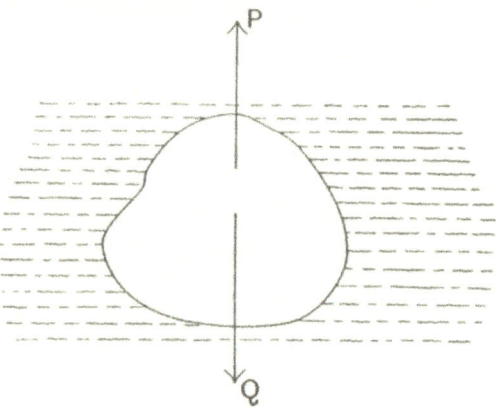

Fig. 6.

And this gives us the reason why a balloon rises
in the air. A cubic metre of air, under standard
conditions of temperature and pressure, weighs 1·293
kilogrammes. A cubic metre of hydrogen gas under
similar conditions weighs only ·088 kilogrammes; if
then we take a light but gas-tight envelope of some
kind and fill it with hydrogen gas, so as to form a
balloon of say 1 c. metre capacity, the balloon will

rise in the air when set free, since there will be a
resultant upward force on it equal to the weight of
1·205 kilogrammes : the balloon could therefore lift
this weight from the ground, and whatever be the
weight of the whole system, envelope, hydrogen,
supports, and car, so long as it does not exceed
1·205 kilogrammes, the balloon will rise until it
reaches a point where the atmosphere is of such a
tenuity that the weight of the displaced air is just
equal to that of the whole balloon system ; it will
then remain in equilibrium.

And now, having raised ourselves into the air, we
have next to ask ourselves how we may direct our
balloon through the subtle fluid in which it is
immersed.

This problem has two aspects—the propelling of
the balloon, and its steering. But it is imperative to
notice that these two problems are intimately connected, in so far that it is utterly impossible to *steer*
the balloon unless we have some means of *propelling*
it—of giving it an *independent* speed through the air.
If you are afloat in a rowing-boat and lose the oars,
then, work the rudder as you will, the boat remains
the sport of every chance current. It is only when
the oars are being used that the rudder has any
effect. And so it is with our balloon. Let us then
turn our attention to the question of propulsion, and
as a simple preliminary illustration, let us consider

the dynamical principles involved in that very common and every-day process called walking. In walking, one presses one's foot hard against the ground ; the earth exerts an equal and opposite pressure on the foot, and this force, which is exerted on the walker, propels him forward. And in just the same way the screw propeller of an airship, whirling round rapidly, beats the air backwards, exerting a backward force upon it. The reaction of the air upon the propeller is, by Newton's third law as we have explained, a forward force which is exerted upon the propeller, and this forward force is transmitted to the body of the airship and urges it forwards.

And now, having seen the principles which govern the raising of a balloon in the air, and its motion through the air, we have to turn our attention to a consideration of the leading principles which govern the motion and flight of aeroplanes—the heavier-than-air machines.

The general laws on which the aeroplane depends were explained just over one hundred years ago by Sir George Cayley. In Jules Verne's and similar stories wonderful airships are described as held up in the air by screws turning about vertical axes. One of the difficulties in realizing this dream is the weight of the engines that would be required to work such propellers. Until an engine has been constructed

giving much more power for its weight than the
lightest yet made, such an airship could not rise
from the ground. Sir George Cayley showed how
by using the screw to drive the airship forward it
could raise and support in the air a much greater
weight than if it were used directly to lift it, a
seemingly paradoxical result.

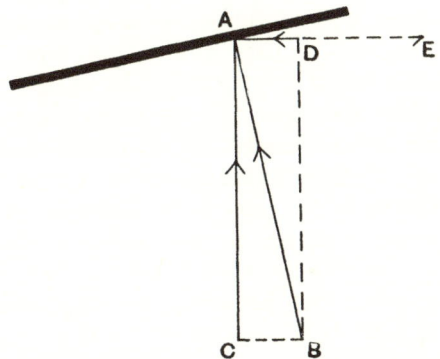

Fig. 7. *AE* direction of motion of inclined plane,
BA air-pressure, *CA* Lift, *DA* Drift.

Suppose a flat surface is driven through the air
horizontally, the front of the surface being raised a
little higher than the rear, so that the under side strikes
the air as it advances (see Fig. 7). The air will exert a
pressure on the under side at right angles to the surface.
This pressure may be considered to consist of two parts

or components, one, vertical, pushing the plane upward called the Lift, the other, horizontal, pushing it back called the Drift. If the surface is nearly horizontal, the front being raised only a little above the back, the air pressure will act almost vertically upward ; so the Lift will be much greater than the Drift. The Lift acting upward supports the weight carried by the plane, while to keep the plane moving the Drift must be counterbalanced by an equal force driving it forward. This force is obtained by the action of the screw propeller. The weight carried, being equal to the Lift, will be many times greater than the propellor thrust which is equal to the Drift, while in a direct-lift machine or *helicoptère* the propeller thrust would have to be the same in amount as the weight. Even so the lightest engines in Sir George Cayley's time and for nearly a hundred years after were too heavy to be successful. The problem before inventors then was to design an aeroplane giving the greatest possible Lift in proportion to Drift and to make it both strong and light so as to be able to carry a heavy engine. The engine itself had to be made as powerful as possible in proportion to its weight. Only by the development, in the first instance in connection with the motor-car industry, of the light-petrol motor has flight by heavier-than-air machines been rendered possible.

All bodies moving through the air have a pressure

exerted on them which, as before, may be resolved into Drift or force resisting motion, and Lift or force pushing the body at right angles to its direction of motion. When there is very little Lift or none at all the Drift, which is then practically the entire air pressure, is called also Head-resistance. Since the weight of an aeroplane is entirely supported by the air, and the engine and propeller of both airships and aeroplanes are necessary to counteract the Drift or Head-resistance, we have next to consider the fundamental principles of air-pressure on bodies moving through the air.

First of all the air-pressure depends only on the rate and direction with which the air and the body meet; it is the same whether the body moves to meet the air or remains still while the air blows against it. This fact is sometimes expressed by saying that a body moving through the air creates an artificial wind. If the air is calm and an aeroplane flies at 40 miles an hour eastward, the air-pressure is the same as if the wind blew at 40 miles an hour from the east, the aeroplane facing it and remaining at rest; or as if the wind were 10 miles an hour from the east while the aeroplane moved eastward at 30 miles an hour; or again as if a west wind of 10 miles an hour were blowing and the aeroplane moved eastward at 50 miles an hour. In all these cases the aeroplane meets the air at a rate of 40 miles an hour.

Obviously the greater the velocity with which the aeroplane and air meet, the greater will be the air-pressure. The weight carried by the aeroplane is equal to the Lift. If the speed through the air increases, the Lift will increase and, becoming greater than the weight, will push the aeroplane upward. If the speed diminishes, the Lift will not be sufficient to support the weight and the aeroplane will begin to come down. For these reasons an aeroplane has a certain speed at which it flies through the air and it cannot be made to fly at different rates, at least not without altering its shape or the position of some parts which at present can only be done to a very small extent. This speed is speed through the air, for on that the air-pressure depends. If an aeroplane's air-speed is 40 miles an hour, then flying with a wind of 30 miles an hour it would have a speed of 70 miles an hour over the ground, and flying against the same wind a speed of only 10 miles an hour over the ground.

How the air-pressure or Drift and Lift on anything varies with the shape of the body, its speed through the air, and the direction in which the air meets it is a question which can only be studied experimentally. Bodies of different shapes and sizes are moved or held in different positions and the velocity with which the air meets them, the Lift and the Drift, are measured. Precautions must be taken to ensure that the velocity and the direction of the motion with which the air and body meet are really known and that these quantities do not change while the experiment is being made. On this account rather elaborate and expensive apparatus is required to make the measurements with convenience and accuracy. Professor Langley was enabled by a grant from the United States Government to carry out a series of experiments, and as full records of these were published they have been of the greatest service to all interested in aviation. In England, Sir Hiram Maxim was in the fortunate position of being able to devote a large sum of money to experimental work, and has published some of his results. The British Government has lately taken up this work and experimental researches are carried out at the National Physical Laboratory.

Three principal methods have been used. Phillips, Langley, Maxim and others employed a long beam, usually called a whirling table, to carry the body round in a horizontal circle. The body may also be held in the centre of a large pipe through which a fan drives a current of air. In this case precautions must be taken to ensure that the air flows steadily and uniformly through the pipe and the body must occupy only a small space in the centre. An apparatus of this sort is in use at the National Physical Laboratory. Experiments have been made also on bodies sliding down wires or along rails, a method which seems to be particularly useful when only Drift or Head-resistance is to be measured. Rougher experiments have been made with bodies attached to engines or motor-cars, or flown as kites or held exposed to the wind.

Experiments in which Lift and Drift or Head-resistance are measured are of primary importance ; but to explain the different effects on bodies of different shapes we must adopt other classes of experiments. In these the disturbance produced in the air is studied. The contribution of each part of the surface to the air-pressure has been measured separately by connecting a small hole in the surface by a tube to a manometer. Bits of thread attached to the surface show how the air passes over it, and some information can be obtained by holding the flame of a candle in the air. Again, if the air is made visible by being mixed with smoke or steam it can be photographed.

Some general principles have now been established. To diminish the Head-resistance, all bodies moving through the air should be as smooth as possible, without projections or sharp edges, shaped so as to part the air gradually in front and allow it to close in again gradually behind. It is found that anything in the way of a sharp edge, a projection, or an abruptly ending stern, causes eddies behind it and greatly increases the Head-resistance (Fig. 8, *B*). It is found best to have a stern gradually tapering to a point for at least two-thirds of the whole length. The bow may be blunter and the widest part should not be more than one-third of the length from the front. 'The head of a cod and the tail of a mackerel' is now

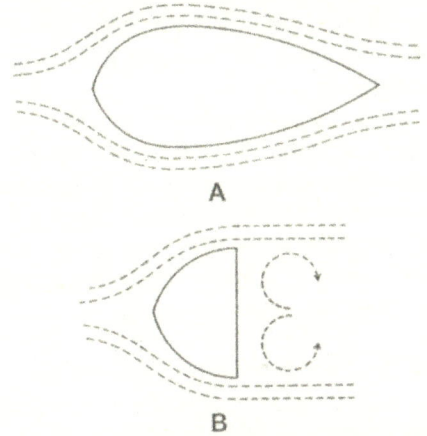

Fig. 8. *A*. Air-current passing over body of approximate stream-line form. *B*. Eddies produced by abrupt stern.

a commonplace expression in books on aviation. This shape is often called the stream-line form, because the air parting in front of the body streams over it and closes up behind without any eddying or any dragging of the air with the body (Fig. 8, *A*). The bodies of fish and of birds are shaped on this plan and closely

2—2

resemble each other. Birds are however broader in proportion to their length than fish, and this is due to the difference in the streamline form for water and for air. This principle is followed in the construction of the gas-bags of dirigible balloons, but the constructors of aeroplanes seem to think that the labour of shaping struts etc. in this way would not be recompensed by the advantage gained in loss of Head-resistance.

Two objects close together experience a much greater Head-resistance than when far apart, and when a number are close together the air in passing through the spaces between them causes a very great Head-resistance. In this case it is better to enclose the objects in a covering which presents a single smooth surface to the air, the covering of course being shaped as far as possible according to the principles enunciated in the preceding paragraph. In some aeroplanes the aviator, engine etc. are enclosed in a boat-like structure of this sort.

The best form for the supporting surfaces of aeroplanes is a very complicated question. First of all we have the principle of Aspect Ratio. When the air produces Lift on a part of the surface it is pressed down by that part of the surface at the same time, and so must have a downward motion communicated to it. This air, since it is now moving in a slightly *downward* direction, cannot be so useful in

producing a Lift or *upward* pressure on that portion
of the surface with which it comes into contact next.
Receiving a further downward motion it will be still
less efficacious on the part of the surface behind that,
and so on. The front portions of the plane, therefore,
produce most Lift, and at a certain distance from the
front a continuation of the surface produces hardly
any Lift at all. For this reason supporting planes
narrow from front to rear with a long front edge give
the best result. The ratio of the breadth from front
to rear to the length parallel to the front edge is called
the Aspect Ratio. In present-day aeroplanes the
Aspect Ratio varies from 1 in 5 to 1 in 8.

To counteract the loss of Lift due to the downward
motion of the air the back of the plane may be bent
down so that the air still presses on it. There is a
limit to the extent to which this may be done. The
air-pressure now is nearer the horizontal, and the
proportion of Drift to Lift is also increased. At a
certain point the disadvantage of increase of Drift
counterbalances the advantage of increase of Lift; so
we are led again to the principle of Aspect Ratio.
The air also has a tendency to escape sidewise at the
ends of the plane, and this loss will be greater the
greater the breadth of the plane.

Surfaces curved from front to rear with the
concavity downward are found to give a larger
Lift in proportion to Drift than flat surfaces, and

consequently the supporting surfaces of aeroplanes are of this form. Phillips was the first to advocate concave surfaces and in 1884 patented some sections which he had found most advantageous. Lilienthal, who is mentioned later in connection with gliders, knew the advantages of concave wings. Different explanations of their efficiency are given.

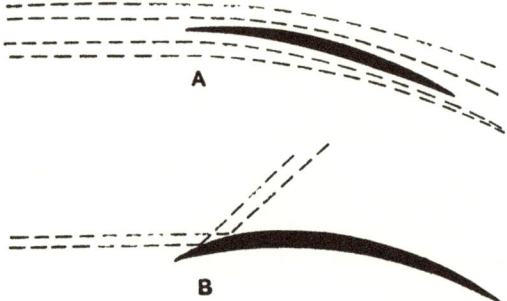

Fig. 9.   *A*.   Gradual deflection of air-current.
     B.   Air-current meeting dipping front edge.

When the air strikes the front edge of a flat surface on the under side, it must receive a somewhat sudden shock, the resulting downward motion accounting for the Lift as previously explained. It is generally held that a better effect is obtained by communicating the downward motion to the air gradually instead of with a shock. This principle is assumed to hold also in the similar cases of windmills, rotatory fans, screw propellers and turbines. According to this explanation, at the front edge the air should pass along the surface tangentially without striking it and be gradually cast down by the curving surface behind (Fig. 9, *A*). The curving of the rear part of the surface has been mentioned before in connection with Aspect Ratio. Phillips on the other hand advocates the 'dipping front edge' (Fig. 9, *B*). He says the curvature should be so great at the front that the air strikes the upper side of the surface and is diverted upwards; this upward motion, according to his theory, leads to diminished pressure on the upper side and consequently to Lift. The form of the upper surface certainly seems to have something to do with the amount of Lift, for a section with the under surface flat and the upper convex gives a good ratio of Lift to Drift. This and many other results cannot be considered as fully explained by either of the theories given above, and further study of the motion given to the air in these cases is evidently required. It seems to be settled by experiment that almost flat surfaces with very small curvatures, give the best ratio of Lift to Drift, while surfaces with a larger curvature give a larger Lift in proportion to the size of the surface, but not so good a ratio of Lift to Drift. Some aeroplane constructors prefer the latter form on account of the smaller surface required for support, though more

engine-power is required to counterbalance the
greater Drift.

In a description of aeroplane wings or supporting
surfaces the edge of the surface which first meets the
air is called the front or *entering edge*.   The edge at
which the air leaves the surface is the *trailing edge*.
A straight line from the front edge to the trailing
edge is called the *chord*.   The length of the surface
in the direction of the front or trailing edge is the

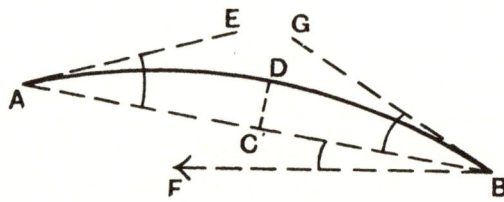

Fig. 10.   *ADB* section of curved surface, *BF* direction of motion,
A entering edge, B trailing edge.

span.   The *Aspect Ratio*, as explained before, is the
ratio of the chord to the span.   The distance of a
point on the surface from the chord is the *camber*
and the position of maximum camber is where the
distance of the surface from the chord is greatest.
The *angle of incidence* is the angle which the chord
makes with the direction in which the plane moves
to meet the air ; the *angle of entry* the angle which
the tangent to the section of the surface at the

entering edge makes with the chord ; and the *angle of trail* is the corresponding angle at the trailing edge. In describing the curvature of the surface from front to rear, the angles of entry and trail, the amount and the position of the maximum camber are generally given.

The wings must be light, and yet strong enough to bear the air-pressure without being deformed. They are made of fabric of some sort stretched on and supported by a framework built up of two main spars in the direction of the span crossed by ribs at frequent intervals. The framework may be covered on one side only, when the wing is termed single-surfaced. To diminish Head-resistance, the spars and ribs in this case are enclosed in pockets of fabric sloping down to the single surface. If the framework is covered on both sides, the wing is termed double-surfaced. In this case the distance between the upper and lower surfaces may be large consistently with good results, and so the main spars may be built-up girders of some depth. These give greater stiffness to the wing and so diminish the number of supporting stays and wires which would otherwise be necessary.

Wenham, as the result of observations on birds flying one above another, propounded the principle that two supporting surfaces, such as are used in aeroplanes, do not materially interfere with each other's action if the vertical distance between them

is not less than the breadth of either. Instead of using
one surface it may be broken up into two or more
placed vertically over each other, a great advantage
when the surface is long. In addition, by connecting
the main spars of two such surfaces by vertical struts
and diagonal tie-wires they can be made to support
each other and the resulting structure is very stiff
and strong. An aeroplane constructed in this way is
called a *biplane.* If there is only one supporting
surface the aeroplane is a *monoplane,* if three a *triplane.* Phillips constructed an apparatus in which
the principles of Aspect Ratio and Superposed Planes
were carried to excess. He used 50 planes each 22
feet by $1\frac{1}{2}$ inches placed 2 inches apart and showed
that a considerable Lift could be obtained.

The shape of the wings of birds may be adduced
in support of these principles. The outspread wings
of birds have a large span in proportion to their
breadth, are concave beneath, and have the steepest
part of their curve close to the entering edge of the
wing. That the body of a bird is so shaped as to
experience least resistance from the air has been
already mentioned. Birds that soar, that is, fly with
outstretched motionless wings and seemingly without
effort, illustrate these principles best, for their flight
presents the greatest resemblance to that of an
aeroplane.

A kite is supported by the air in the same way as

an aeroplane. The kite is held at rest and the air blows against it, but the effect produced is exactly the same as if the kite moved to meet the air and the air remained at rest. Lift and Drift, then, are produced by the air and in this case are counter-balanced by the pull of the string on the kite. The larger the Lift in proportion to the Drift, the nearer to the vertical will be the string ; while the smaller the Lift in proportion to the Drift the more nearly horizontal will be the string where it is fastened to the kite. It is easier to construct a kite that will fly than an aeroplane, because the ratio of Drift to Lift does not require to be so small, but the Lift of course must be large enough to raise the weight of the kite and some string as well. The string of a kite may be regarded as taking the place of the engine and propeller of an aeroplane as regards weight and propeller thrust, but without the limitation to which the propeller thrust to be obtained from an engine of given weight is subject. That kite which has the greatest Lift in proportion to Drift will rise highest and nearest the vertical, and so the construction of a good kite and of an aeroplane are very similar problems.

CHAPTER II

PROPELLERS AND MOTORS

§ 1. *Propellers*

THE principle of action and reaction states that a force, tending to move a body in one direction, must have corresponding to it a force tending to move some other body in the opposite direction. An airship or aeroplane is completely surrounded by air ; in consequence the only other body on which the reaction, corresponding to the force propelling the airship or aeroplane forward, can act is the air itself. The aeroplane is driven forward and the air backward at the same time. To drive the air back a screw propeller is used.

Any screw, that of an office press or of a lifting jack for example, consists of a central axis with a projection called the thread running spirally round it. The screw fits into a nut in which a groove has been cut corresponding to the thread of the screw. Turning the screw round causes the thread to press against this groove and produces a force tending to

move the screw along its axis, and the nut in the
opposite direction. Supposing the nut to be fixed
when the screw is turned round, the thread runs
along the groove, and the screw moves forward.
When the screw has made one complete turn, the
distance which it has moved forward is called the
pitch of the screw. In any case, whether the screw
or the nut, or both, move, the distance the screw
moves forward added to the distance the nut moves
back during one turn of the screw is the pitch. The
ordinary screw used by carpenters works in this way,
the thread of the screw cutting its own groove in
the wood. If we suppose the axis of the screw to
be small and the thread to project a long way from
the axis, and we cut a slice from the screw, the slice
consists of a circular disc with a long projecting arm
which is the part of the thread cut off. This represents
roughly the sort of screw employed for propellers,
but screws with two threads at least are used, so that
the propeller has two or more projecting arms. These
latter are called blades and form short parts of the
threads of the screw.

The nut in which the propeller of an airship works
is the air. If the air remained steady, while the
propeller turns round once, the airship would move
forward a distance equal to the pitch, and its speed
per minute would be the pitch multiplied by the
number of revolutions per minute made by the screw.

The air however cannot remain still, for the pressure of the propeller blades casts it back. As a result the air is sucked in from the front and all round. Evidently the propeller cannot cast back the air any faster than it could a solid nut. So if we suppose the air after being cast back to move as a fixed nut would, we take the most favourable view possible of the action of the propeller. This would make the pitch of the screw equal to the distance the screw moves forward added to the distance the air moves back during one revolution. Stated in another way, the forward velocity of the propeller added to the backward velocity communicated to the air is equal to the velocity with which the propeller would move turning at the same rate in a fixed nut. The velocity communicated to the air is called the slip of the screw. It is clear therefore that the pitch of a screw propeller must be greater than a certain amount to be of any use at all. For example, an aeroplane moves at the rate of 45 miles an hour, that is, 3960 feet in a minute. If the propeller makes 1000 revolutions a minute, the aeroplane moves forward nearly four feet while the propeller makes one revolution, and consequently the pitch of the screw must be more than four feet. At the same time the pitch must not be too great ; if it were, the blades would stir the air round instead of casting it back. For each particular speed of the airship and speed of revolution of the propeller there will be one pitch which will give the best result, and this can only be determined experimentally.

The revolving blades trace out a circle whose diameter is called the diameter of the propeller, and the air they reach is set in motion. Again, taking the most favourable supposition, the air cast back is a cylindrical column whose cross-section is the circle referred to. Now the principles of mechanics show that the force, driving the aeroplane forward and the air back, is proportional to the quantity of air cast back in any time multiplied by the velocity given to it. The work done in casting the air back, and consequently the amount of the power of the engine expended in moving the air, is proportional to the same quantity of air multiplied by the square of the velocity. Now this part of the power of the engine, usually said to be expended in slip, is wasted as far as driving the aeroplane is concerned, and should be made as small as possible. It varies as the square of the slip, so the slip should be made as small as possible. To keep up the propelling force the quantity of air cast back must be made large, and so the diameter of the propeller is large. Propellers of large diameter and small slip are, therefore, most economical of power. If the diameter of the propeller is made smaller, the slip must be increased to produce the same thrust. As already pointed out, this cannot be done with advantage by increasing the pitch of the screw, leaving its rate of revolution unaltered. The speed of revolution must be increased. To produce the same thrust propellers of large diameter and slow speed of revolution may be used, or propellers of small diameter revolving more rapidly. The former have the smaller slip and are more economical of power.

This is, however, not the only consideration to be taken into account. Large propellers are heavier, and the weight puts a limit on the size of propeller that can be used. Again, if the speed of revolution of the propeller differs from that of the motor, the two must be linked up by some system of gearing by which the power of the motor is transmitted. In this transmission a certain amount of power is lost. For each size of propeller the power lost in slip, the power lost in gearing, and the disadvantage of weight and size must be estimated, and that propeller employed which is on the whole most advantageous. The petrol motor used on airships and aeroplanes has a very high speed of revolution, over 1000 revolutions per minute as a rule. If the propeller is mounted directly on the crank-shaft of the motor, it has a small diameter; thus a considerable amount of power is lost in slip but none in gearing. Such propellers are employed on almost all aeroplanes and on some airships. A few aeroplanes and some airships have larger propellers connected to the motor by gearing.

The blade of a propeller is not usually a simple screw of the same pitch throughout. A better effect may be obtained by varying the pitch as the distance from the axis increases, and also varying it across the blade from front edge to back. Experiment is here the only guide. By experiment, too, the best breadth of the blade must be found as well as the number of blades to be used. Attention may be called to some important points connected with the testing of screws. The explanation given above shows how much the motion of the airship through the air has to do with the effect of the propeller. While being tested, the propeller should move through the air at the speed of the airship. As in testing supporting planes etc., two methods are available: the propeller may be mounted, at the end of a rotating arm for example, so as to move through the air; or the propeller, remaining still, may act in a current of air blowing at the same speed in the opposite direction. Many tests on model propellers have been made. A large rotating arm has recently been erected at Barrow on which propellers of the largest size may be tested.

Makers of propellers in advertising their wares have often published statements of the thrusts obtained at a certain number of revolutions per minute. These statements were quite valueless for two reasons. In the first place the tests were made with propellers at rest in still air. In the next place, the thrust at

a certain speed of revolution is a matter of no importance ; a very large thrust, if obtained by using
a very powerful and heavy motor, is no advantage.
The thrust per horse-power of the motor used is the
important thing, according to which the merit of the
propeller ought to be estimated.

It has been said that the best form of blade in
a propeller is a matter for experiment. Some general
principles must be adhered to, very similar to those
which govern the form of the supporting surfaces of
aeroplanes. The blade of the propeller is to move
through the air casting it back, instead of down ;
and the air-pressure on it can be resolved into two
components, one parallel to the axis and one perpendicular to it. The former is the useful component,
pushing the propeller forward ; the latter is not
useful, but resists the motion of the blade and has
to be counteracted by the action of the motor, and
therefore its ratio to the useful component has to
be made as small as possible. This is almost the
same as the problem of Lift and Drift, and the same
general principles apply. The fact that the propeller
blade moves faster at the tip than nearer the axis,
alone distinguishes the propeller blade from the
supporting surface. First, suppose the blade shaped
of such a pitch that it will move through the air
without disturbing it, as if the air were a fixed nut.
In this case, like a horizontal surface moving edge-

wise through the air, it meets the air edgewise and
experiences no thrust. Now alter the shape of the
blade so that the air strikes it on the back and is
deflected backwards. In accordance with the principle
of Aspect Ratio the blades should not be too broad ;
in accordance with that of Superposed Planes there
should not be too many blades, otherwise they interfere with one another's action. The blades should be
concave at the back and convex at the front with
both surfaces smooth. The principle of the dipping
front edge or the principle of the gradual deflection
of the air without shock will apply equally to the
propeller blade and to the supporting surface of the
aeroplane.

Propellers are made of both wood and metal.
Wooden propellers are built up of strips of wood
running from the axis to the tip of the blade. The
strips are first fastened together and afterwards
fashioned to the desired shape. In metal propellers
the blade is formed of a sheet of metal supported
by a rib running out from the axis.

In ships the propeller is always placed behind the
ship, never in front. There are two reasons for this.
First, if the propeller were in front, the water thrown
back would strike the ship and impede its progress.
Secondly, on account of the friction against the sides,
the water at the stern is dragged forward to some
extent, with the ship ; the propeller if placed behind

has first of all to destroy the forward motion of the
water, before it can give it a backward velocity.
This velocity is then smaller and the work expended
in slip is therefore less, and consequently the propeller
acts more efficiently.   The same considerations apply
in the air, but are of less importance and are generally
overridden for constructional reasons in airships and
aeroplanes. The propellers are usually placed behind
the main planes in biplanes, but in monoplanes on
account of the presence of the boat-shaped body
which carries the tail the more convenient position
is in front.

§ 2. *Motors*

The importance of a light and efficient motor
in aviation has been emphasized elsewhere.   For
a long time the steam engine was the only available
motor, and many attempts were made to construct
a light steam engine. Langley and Maxim each
achieved a certain degree of success, the former on a
small scale, the latter on a large one.   In both cases,
however, supplies of fuel and water for short flights
only could be carried. The difficulties which have
made complete success impossible arise from the
very nature of the steam engine. The steam has a
temperature very much lower than that produced by
the combustion of the fuel, therefore a large part of

the energy of the fuel is necessarily wasted in passing to the steam and a large supply has to be carried. Further, if a non-condensing engine is employed, as is done on railways, a large supply of water is required for a long flight, while if a condensing engine is employed, so that the same water can be used over and over again, the weight of the condenser and the water in it is considerable. The boiler also adds to the weight. The development of the internal combustion motor in connection with the motor-car industry opened up new possibilities. For in it no boiler or water for conversion into steam is required. The gases produced by the combustion of the fuel are used to drive the piston. These gases are at so high a temperature that they would destroy any pipes or valves they passed through; the combustion therefore takes place in the cylinder itself. The fuel is converted into gas, mixed with air in the correct proportions for complete combustion which takes place instantaneously, i.e. as an explosion, when the mixture is fired by an electric spark. This direct use of the products of combustion is the chief cause of the superiority of the internal combustion motor over the steam engine. The gas is cooled by its own expansion as it drives back the piston and is able to escape from the cylinder without destroying the valve.

The internal combustion motor is usually a four

cycle one, that is, work is done by the piston only during one stroke in every four. Suppose at the beginning of a down-stroke an explosion takes place, the piston is driven down and work is done. During the next up-stroke, the exhaust stroke, an exhaust valve is opened through which the piston drives the gas that has done the work. This valve is then closed. A down-stroke, termed the admission stroke, follows, during which an admission valve is opened and the mixture of vaporised fuel and air sucked into the cylinder. Then all the valves are closed and in the next up-stroke (the compression stroke) the mixture admitted is compressed at the upper end of the cylinder. Now the mixture is ignited by an electric spark, and another useful stroke follows. To keep the engine working steadily, if it has only one cylinder, a heavy flywheel is used. This acts as a reservoir of power, storing up energy during the useful stroke and giving it out again to keep the engine running during the three idle strokes. The use of a large number of cylinders enables the flywheel to be dispensed with or at least reduced in size. The cylinders ought to be arranged so that their useful strokes begin successively at equal intervals. The different moving parts should be balanced to avoid vibration, the motion of one part in one direction being counterbalanced by the motion of other parts in an opposite direction. For example, with two cylinders their pistons should always be moving in opposite directions. Placing the two on opposite sides of the crank shaft, the advantages of balance and of working strokes at equal intervals can be combined, every alternate stroke being a useful one on the part of one of the cylinders, but half the strokes would still be idle ones for both cylinders. At least four cylinders must be used to make their useful strokes continuous, one beginning when another stops, and eight cylinders are common in aeroplane motors.

The fuel used is petrol which contains the lighter constituents of mineral oil and vaporises readily at ordinary temperatures. Sometimes the vaporising and mixing are done in the cylinder, the petrol being sprayed in by a pump. The mixing however is not well done by this method. More usually these operations take place in a separate vessel called a carburetter. This must work regularly, always supplying to the cylinders air and petrol vapour mixed thoroughly and in the correct proportions.

The valves must be reliable. They must be strong enough to remain closed during the explosion. The admission valve must admit a sufficient quantity of the mixture, and the exhaust valve must allow the spent gases to escape readily without unduly impeding the motion of the piston. These valves only open during the alternate down- and up-strokes respectively ; accordingly a separate shaft, called the cam shaft, has to be provided to operate them, geared so as to revolve once for two revolutions of the crank-shaft.

The construction of a reliable apparatus for producing the electric spark to ignite the mixture is a separate problem.

The cylinder, piston, and valves have to be kept cool, for the expansion of the gases cannot be carried so far as to prevent these parts from being unduly heated. The cooling is effected by passing over them a current of either air or water. In aeroplanes the passage of air over the cylinders as the aeroplane moves through the air is sometimes relied on. This method seems to be efficient for a short period only, unless, as in the Gnome engine, to be referred to later, the cylinders themselves revolve. Water cooling is more general. A jacket encloses the cylinder and water circulates through the narrow space between the jacket and the cylinder wall. This water passes to a radiator where it gives out its heat to the air and returns cool to the cylinder to go round the same circuit again.

The parts of a petrol motor move very rapidly ; the crank-shaft usually makes more than 1000 revolutions per minute. This rapid motion would soon heat and destroy two parts bearing on each other unless they are well lubricated so as to banish friction. A system by which lubricating oil is automatically conveyed to all bearing surfaces is therefore necessary.

Motor-car engines have by years of successive improvements been made very reliable. For airships engines of the same sort are generally employed, extreme lightness not being so important as reliability. Such engines however are too heavy for aeroplanes, so special engines, lighter in proportion to their power, have to be constructed. To reduce the weight, materials very strong and light may be used regardless of their cost or of the expense of working them. All parts not absolutely necessary may be dispensed with and the weights of all other parts cut down until very little margin of safety is left. Many aeroplane engines are only motor-car engines specially lightened in this way—a method which gains lightness at the expense of reliability. A better way to attempt the solution of the problem is to construct an engine of a different pattern, specially suited to aeroplane work. A radial arrangement of the cylinders, i.e. an arrangement round the crank-shaft like that of the spokes of a wheel round the hub, seems the most suitable. All the pistons here may be connected to the same crank— an arrangement which saves a great deal of weight. If the number of cylinders is odd and the explosions occur in succession in alternate cylinders, first, then third, then fifth, and so on, the intervals between successive explosions will be equal. In many engines a partial radial arrangement of the cylinders has been adopted but cooling has been the chief difficulty to overcome. Sometimes the cylinders revolve while the crank-shaft remains fixed, the action of the pistons remaining the same. The rotation of the cylinders through the air keeps them cool. Centrifugal force of course acts on all rotating parts, and with rotating cylinders different arrangements have to be made for admission and exhaustion and for lubrication. The mixture of petrol vapour and air has sometimes been supplied through the hollow crank-shaft and piston rod, the admission valve being placed in the piston. Motors of the rotary type have the disadvantage that a considerable amount of power is wasted in whirling the cylinders through the air. However, one of them, the Gnome, is generally regarded as the most powerful and reliable aeroplane engine of the present day. Though it does not now hold the record for length of flight, more long flights have been made with it than with any other engine. Lubrication seems to be a difficulty, and after a day's flying it is said to require taking to pieces completely for cleaning.

Suggestions have been made for a non-revolving radial engine, cooled by being placed in the current of air produced by the propeller.

Anyone reading accounts of flights and of aviation meetings in the press soon sees that aeroplane motors are very unreliable. The ignition, the carburetter, the lubrication, the cooling, and the valves all give trouble. The failure of any of these, if it does not stop the engine altogether, causes its power to fall off and very probably brings the flight to an end. In fact the perfect working of a motor for more than a short time is sometimes mentioned as if it were a piece of extraordinary good luck. Reliability trials are very urgently needed. Mr P. Y. Alexander offered a prize for a British engine which could run unattended for 24 hours and fulfil certain other conditions. The results of the tests in this competition have recently been announced. Only one engine, the Green, ran for the twenty-four hours but it did not yield the full power required. Of other British engines long flights have been made by the E. N.V. motor towards the close of 1910. The reliability runs which aided so much in the development of motor-car engines may be recalled. In these all breakdowns of the engine were counted against it. Similar trials in which an aeroplane makes flights on successive days, the repairing and overhauling of the motor being either forbidden or carried out under official observation, ought to lead to great improvements in aeroplane motors. Such trials are needed before the aeroplane can become a vehicle for either popular or military use. At present an aviator has to employ a staff of mechanics and keep a reserve of engines and spare parts, and repairs and overhauling are carried out after almost every flight.

If the operations of exhaustion and admission can be carried out effectively during a single stroke of the piston, the number of working strokes can be doubled. The engine becomes a two-cycle one, and each cylinder in it does as much work as two in a four-cycle engine. Other more visionary developments depend on the discovery of some metal or alloy which would not be destroyed by the very hot gases resulting from the combustion of the fuel. If such materials can be found, the combustion can take place outside the cylinder, the gases passing into the cylinder as steam does in a steam engine. A motor turbine even could be constructed to be actuated by these gases. An electric accumulator, light enough to be a practicable source of power for aeroplanes, is not to be expected in the immediate future.

CHAPTER III

§ 1. *Stability*

ALTHOUGH an aeroplane is balanced correctly, it may fly only a short distance. After some time it may overturn in either of two ways—tilting head up or down, or canting sideways. The upsetting may be completed in a continuous swerve of the aeroplane from its correct position in flight, or may be preceded by pitching or rolling motions which become more and more violent. This tendency of the aeroplane is not due to any fault of balance and cannot be cured by any adjustment of the weight. An aeroplane which flies steadily, without showing this tendency to upset, is said to possess stability. It has longitudinal stability when it has no tendency to upset by tilting head up or down, and lateral stability when it has no tendency to upset by canting sideways.

There are three ways of obtaining stability. The

aeroplane may be so designed that it returns of itself to its correct position after any departure from it. This is called inherent stability. Or the pilot may bring the aeroplane back to its correct position ; in this case stability is obtained by the method of personal control, and the problems of stability and steering become identical. The third kind, automatic stability, is a development of the method of personal control, the mechanism which corrects a deviation being put in operation automatically instead of by the pilot. Of these three, inherent stability is obviously the most desirable if there are no countervailing disadvantages attached to it. No attention on the part of the pilot is required and there is no mechanism to get out of order. Inherent longitudinal stability is easily obtained, but the problem of inherent lateral stability is more complicated.

All that is required to ensure inherent longitudinal stability is a surface placed horizontally behind the aeroplane so as to meet the air edgewise, when the aeroplane is flying in its correct position (Fig. 11, *A*). This surface is called the horizontal tail. Normally the air passes over the upper and lower surfaces of the tail without striking it. If the head of the aeroplane tilts downward, the air strikes the upper surface of the tail and raises the head again (Fig. 11, *B*). If the head tilts up, it is brought down in a similar way. This action of the tail depends both on its size and

on its distance behind the main planes. To increase
its effect either of these two may be increased. Of
course the effect of a tail may be too small to keep
the aeroplane steady : the early gliders seem to have
made the mistake of using tails which were too small
and too close to the main portion of the aeroplane. This principle, like many of those mentioned
previously, is borrowed from birds.

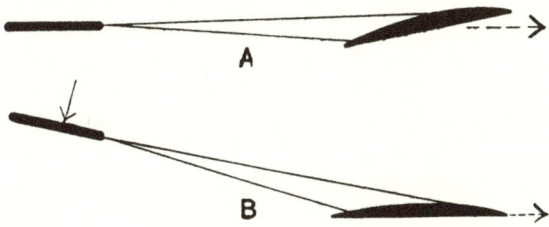

Fig. 11. *A*. Aeroplane with neutral tail in normal flight.
B. Aeroplane righted by air-pressure on tail.

A tail which meets the air edgewise may be called
a neutral or non-lifting tail. It is not absolutely
necessary that the tail should be neutral; the essential
point seems to be that there should be two surfaces,
of which the front one meets the air at a greater
angle than the rear one ; in other words the former
surface experiences a greater Lift in proportion to
its area than the latter. Many aeroplanes have a
lifting tail. In some the tail is even set parallel to

the main planes. Here the stability is properly
explained by the fact that the air is cast down by
the main planes and so meets the tail at a smaller
angle. In some aeroplanes the rear surface is the
larger and forms the main supporting surface, the
smaller surface in front being carried tilted up to
meet the air at a greater angle. On the whole the
neutral or non-lifting type of tail seems to be best.
Hargrave kites, however, and the Voisin type of
biplanes which have been developed from them, have
large tails set parallel to the main planes and both
possess great stability. In all successful aeroplanes
at the present time the principle of inherent longitudinal stability is adopted. Even the Wrights have
abandoned their early practice and now fit horizontal
tails.

The ease with which the aeroplane can be turned
round must be taken into account in determining the
size and position of the horizontal tail necessary for
stability. Take a rod and a weight which can be
attached to the rod at different positions along it.
Whirl the rod and the attached weight round in a
circle. The further the weight is from the centre of
the circle, the greater is the effort required to start
the rod whirling. Again divide the weight into two
equal parts and place them some distance apart on
the rod. It will be found that it is harder to start
the rod and weights whirling than if the two weights

are placed together midway between their former positions. The same principle is exemplified in the flywheel of an engine. By placing as much of the weight as possible in the rim, the difficulty of starting or stopping the wheel is increased. A body which offers a great resistance to being turned about an axis is said to have a large moment of inertia about that axis. On this principle in an aeroplane the less the difference either in height or longitudinally (i.e. behind or in front of each other) between the positions of the different parts, the easier it is for the tail to turn the aeroplane round, and the better the stability. To ensure longitudinal stability it is best to place the different parts as far as possible side by side and at the same height. This is the arrangement adopted in the Wright aeroplane, the pilot, passenger and engine being placed side by side. Most other constructors however place the pilot, engine, etc. in a line from front to rear, an advantageous arrangement from the point of view of Head-resistance and the saving of engine power.

The problem of inherent lateral stability is very difficult to solve. Suppose an aeroplane has acquired a sideways cant. The forces acting on it are the weight, the propeller thrust, and the air-pressure. This last is the same in amount before and after canting, but its direction turns round with the aeroplane. It is clear that none of these forces will right the cant.

The combined effect of the weight and the Lift, which
having turned with the aeroplane does not now act
vertically, is to make the aeroplane move in a sideways and downward direction towards the depressed
wing-tip.   Only after this sideways motion has begun,
can an air-pressure be brought into play to right the
cant.   This fact is often urged against dependence
on inherent lateral stability.   The recovery in this
way, it is said, must be slow, and the accompanying
sideways motion is objectionable as it takes the
aeroplane out of its course.   These are rather reasons
for supplementing the inherent stability by other
devices which would correct the cant more rapidly.
The inherent stability should be retained for the sake
of safety and to allow the pilot to devote his attention
for a time to other matters.

To produce inherent lateral stability the sidewise
motion set up by a cant must exert an air-pressure
on some part of the aeroplane.   This at once suggests
the use of fins, or vertical surfaces, in suitable positions,
meeting the air edgewise when there is no motion
sideways.   By an adaptation of the principle of
superposed planes two or more fins placed side by
side at sufficient distances apart may be used instead
of one large fin.   At first it seems obvious that a fin
placed above the aeroplane would raise the depressed
wing-tip.   In such a position, or in fact in any position,
it is very doubtful if one fin will ensure stability.

The action of the fin has to be sufficient to right the aeroplane and not great enough to cant it still more towards the other side. It seems better to use two fins. One may be placed in front and one behind, either level with or slightly above the main portion of the aeroplane, certain relations being maintained of course between the sizes and positions of the fins. Another possible combination would be one fin above and one behind, the fin behind being more than a certain distance from the main plane.

Instead of vertical fins other devices may be used. The fin above may be replaced by two surfaces to which the name stabilisors may be given. These would be placed respectively far to the right and left of the aeroplane near the wing-tips, inclined upwards from the centre of the aeroplane, and meeting the air edgewise in normal flight. These, in conjunction with a fin at a proper distance behind, should produce lateral stability.

Aeroplane wings set at a dihedral angle produce an effect similar to that of a fin above or a pair of stabilisors. The wings are inclined so that their tips are a little higher than their centre, forming thus a very much flattened V when viewed from the front or rear (see illustration of Antoinette monoplane, page 87).

This principle, which is often used in practice, seems to be very efficient in producing lateral stability when adopted in conjunction with a fin behind. But it is open to the serious objection that it causes a waste of engine power. The Lift on each wing is partially wasted, for it does not act vertically upward owing to the inclination of the wing. To compensate for this the wings must be made larger; this increases the Drift, which again necessitates increased propeller thrust and increased engine power.

Another device in connection with lateral stability was first introduced by Mr Weiss and has lately proved successful in the aeroplane constructed by Lieutenant Dunne. Here the wings extend horizontally from side to side, but are twisted so that the Lift is greatest at the centre and diminishes gradually until, near the tips, it vanishes altogether. At the tips it is said to diminish still more, that is, it becomes negative and pushes the wing down instead of up. This variation of the Lift is obtained by altering the form of the cross-section of the wing from the front edge to the trailing edge. Very probably the best form for the wings of an aeroplane has not yet been discovered. Two things can be varied at different distances along the wing—the cross-section and the inclination of the wing to a horizontal line drawn from one wing-tip to the other. Horizontal twisted wings, wings set at a dihedral angle, and horizontal wings with stabilisors fitted as extensions, are all particular ways of making these variations. Many other combinations might be devised and tested.

Whatever form of wings is used, it is probable that a vertical fin in some position would also be required to give good lateral stability.

In Hargrave kites and the original Voisin biplanes vertical panels between the main planes and between the planes of the biplane tail are utilised to give lateral stability. In kites and gliders they act satisfactorily, but aeroplane constructors seem to have decided that their disadvantages outweigh their advantages.

One of the objections made to inherent lateral stability has been mentioned. Another disadvantage associated with inherent stability, both longitudinal and lateral, is the fact that variations in the wind must cause pitching and rolling. If the wind alters its direction so as to blow more upwards or downwards, the same effect is produced on the tail as if the aeroplane tilted. The aeroplane then must move as a ship does in pitching. If the wind veers round so as to strike the aeroplane on one side or the other, the same air-pressure is produced as if the aeroplane moved sideways, and the inherent lateral stability must make it cant or roll. With this rolling and pitching are associated the difficulty and danger of flying in a high wind. Flying in a wind, as already stated, is exactly the same as flying in calm air; but this needs the qualification that the wind referred to is perfectly steady, always blowing from the same

direction at the same rate. In practice the wind
never blows in this way, it always comes in gusts
and also changes its direction. In addition it varies
in different places, especially when diverted by
obstacles or inequalities in the ground. In general
the higher the wind, the greater its variations and
the greater the difficulty of flying in it.

An aeroplane may be upset by a sudden squall
as a sailing ship may be capsized, or a succession of
gusts may make it pitch or roll more and more until
at last it overturns. Ships have been lost in this
way, rolling dangerously under the influence of
certain waves and at last overturning. The designers
of the ships had not taken account of this source of
danger; but when the problem was investigated it was
found easy to design ships with inherent stability
which would not roll excessively under the action
of any waves they would ever meet with, and would
right themselves even after heeling very far over. It
is possible that aeroplanes could be designed to fulfil
the same conditions, both longitudinal and lateral
stability, i.e. both pitching and rolling, being taken
into account.

The study of these problems is much more complicated in the case of the aeroplane than in that of
the ship. Take the problem of longitudinal stability
and pitching. When the aeroplane pitches, the air
meets the aeroplane at a different angle so that the

Lift and the Drift must change. The alteration in the
Lift makes the aeroplane rise or fall, and the alteration in the Drift changes its velocity. In fact any
one of the three things—the pitching motion, the
change in the direction of motion, and the change
of velocity—must influence the other two. A complicated problem such as this cannot satisfactorily
be investigated without the use of rather advanced
Mathematics. This does not mean that the problem
can be solved by Mathematics alone, but that
Mathematics is absolutely necessary in order that
experiments may be intelligently devised and that
the correct deductions may be drawn from them.
Some well-known writers on Aviation decry Mathematics, but do not seem to be aware that Mathematics
is only a powerful means of reasoning and that
Mathematicians, like other people, are unlikely to
reach right conclusions, if they start from wrong
premises.

Important advances in the mathematical discussion
of aeroplane stability were initiated by Professor
G. H. Bryan and Mr W. E. Williams at Bangor in
1903. Mr Williams has also studied the question of
stability experimentally. He photographed the actual
paths of model gliders through the air by attaching
magnesium wire to them. By placing a rotating wheel
in front of the plate he was also able to measure the
velocity of the gliders. The trace of the path is

divided by means of the wheel into a series of dashes with intervening spaces ; by measuring these dashes and gaps the velocity can be estimated.

The problem of lateral stability and rolling is similarly complicated. As explained before, a rolling or canting motion is accompanied by a motion of the aeroplane from side to side, which in its turn will generally cause the head of the aeroplane to veer round. Here again we have three motions influencing one another and the problem is even more complicated.

The Wrights were fully aware of the danger to an aeroplane possessing inherent lateral or longitudinal stability of being overturned by a sudden squall. They also considered any rolling or pitching undesirable, and for this and other reasons decided on complete absence of inherent stability. They aimed at making their aeroplane uninfluenced by gusts of wind so far as pitching and rolling were concerned, and relied exclusively on the action of the pilot to restore it to its correct position after tilting or canting. This system necessitates great skill and consequently a long period of training on the part of the pilot, and during flight his constant attention is demanded. These must be considered disadvantages so far as the popular use of aeroplanes is concerned. The less the skill required on the part of the pilot and the more attention he can give to other things,—his engine, for example, or the country over which he is passing,—the more popular will aeroplanes become. Even so far as flying in gusty weather is concerned, the Wright aeroplanes do not seem to have any advantage over others. The Antoinette monoplane is generally credited with the best performances in high winds, and it possesses inherent stability both longitudinal and lateral. Though it is impossible to say how much of the credit for these flights is due to the design of the aeroplane and how much to the skill of the pilot, it is certain that Antoinette aeroplanes with their inherent stability are better adapted for flight in gusty winds than Wright aeroplanes with their absence of stability. Recently, as stated elsewhere, the Wright brothers have departed from their principles with regard to longitudinal stability.

An aeroplane for general use should possess inherent longitudinal and lateral stability. The range of stability should be as large as possible, that is the aeroplane should right itself even after tilting or canting far over. The means employed to give stability should not cause excessive rolling or pitching in any winds likely to be encountered, nor should they permit the overturning of the aeroplane by a sudden squall. Whether all these requirements can be satisfied is a matter for investigation. The inherent stability should be supplemented by a system of personal or automatic control, especially in the case of lateral stability. The control system would be especially useful near the ground, in the neighbourhood of any obstacle or of other aircraft, while the inherent stability would ensure safety when high above the ground with nothing in the vicinity.

There is a third sort of stability—directional, which keeps the aeroplane's head directed towards the same point of the compass. This sort of stability is inseparable from lateral. It is impossible to ensure complete inherent directional stability. If the head of an aeroplane has been turned by some accident towards a different point of the compass, no air-pressure can be exerted which will turn it back to its old direction. The lateral and directional stability possessed may be such that after interference by a gust of wind or some such accident, the aeroplane will settle down into steady motion as before but towards a different point of the compass.

§ 2.   *Control*

The three different ways in which an aeroplane can turn round have been described in connection with longitudinal, lateral, and directional stability respectively. For purposes of steering and of bringing the aeroplane back to its correct position after any deviation, three controls are necessary by which the pilot can turn the aeroplane about a transverse horizontal axis, a fore and aft axis, and a vertical axis respectively. The three corresponding controls may be called longitudinal or vertical control, lateral control, and directional or horizontal control.

For longitudinal control a horizontal rudder, usually called an elevator, is required. This is placed either before or behind. In the latter case it may be identical with, or form part of the tail which gives longitudinal stability. When not in use it meets the air edgewise, and the pilot by a system of wires and levers, can raise or depress the rear edge so that the air strikes the upper or under surface as required. If the elevator is in front, the head is tilted down or up respectively by the air pressure on the elevator, while if the elevator is behind these directions are reversed. It must be remembered that the elevator cannot be used alone to direct the course of the aeroplane up or down. Its use is limited to turning the head in the required direction, and it should be restored to its normal position when this has been accomplished. For a short time under the influence of its previous velocity the aeroplane will move in the direction in which its head is turned; consequently the elevator can be used to make the aeroplane take a jump. The path which will ultimately be followed, depends on the balance of the aeroplane and on the mutual relations of the weight, the propelling force, the Lift and the Drift, the latter two depending on

the velocity. To make the aeroplane ascend, its
head might be turned up by means of the elevator,
and at the same time the engine accelerated, for
more power is required to drive the aeroplane up
hill. This upward motion might be obtained, without
the use of the elevator, by engine acceleration alone
if the stability is sufficient. If the propeller thrust is
increased, as already explained, the velocity must
increase ; as a result the aeroplane will rise in
the air on account of the increase in the Lift. The
same considerations apply to the descent, the propeller thrust being diminished instead of increased.

If the elevator is carried normally in a different
position during flight, all the conditions of flight are
altered. The balance, the direction in which the
aeroplane meets the air, the velocity required for
support, and the propeller thrust necessary for motion
in any path, are all affected. By making the main
surfaces themselves adjustable or by detaching them
and substituting others, greater alterations in these
conditions may be made.

The problems of lateral and directional control
are linked up together. When one side of the
aeroplane is to be raised, an increased Lift is obtained
on that side with or without a diminution of the Lift
on the opposite side. In general an increase of Drift
will accompany the increase of Lift, the wing which
is raised will also be retarded, and the direction in

which the aeroplane is heading will be altered. The directional control must be used to correct this. Again, when the aeroplane is flying round a curve the wing towards the outside of the curve moves faster than the wing on the inside, and the Lift on the former must be greater than that on the latter. The wing on the outside of the curve is consequently raised, and to correct this, or rather, as we shall see, to limit the rise to the correct amount, the lateral control must be used.

Wing-warping is the mode of lateral control most in use. The front edge of the wing remains rigid, while the rear edge near the wing-tip is pulled a little down or up. The lowering of the rear edge increases the angle at which the air meets the wing and consequently increases both the Lift and Drift on it. The raising of the rear edge diminishes both the Lift and Drift. The motions are usually controlled by a single lever or wheel. Ailerons may be used instead of wing-warping. These are small supplementary planes carried far to the right and left near the wing-tips. When not in use they are horizontal and meet the air edgewise ; by depressing or elevating the rear edge the same effect is produced as in wing-warping. Theoretically ailerons have the advantage. If they are linked up so that the rear edge of one is depressed exactly as much as the rear edge of the other is raised, the Drifts on the respective

ailerons will be equal and there will be no tendency
to turn the head of the aeroplane round. In practice, it
is stated, the rudder has to be used on account of the
difficulty of adjusting the ailerons to obtain this result.

For directional control a vertical rudder is placed
either before or behind the aeroplane. This is moved
and acts in exactly the same way as the rudder of
a ship. It may form part of or be identical with one
of the fins used to give lateral and directional stability.

Each of the three controls may be actuated by a
separate lever or wheel, one operated by each hand
and one by the feet. More frequently two and
sometimes all three are linked up to a single lever.
A lever, supposed for the sake of simplicity to be
vertical, may be mounted so as to be capable of four
different motions—first, a backward and forward
motion about a pivot ; second, a similar movement
from side to side; third, a sliding movement up and
down through a collar ; and fourth, a rotation also in
a collar, produced either by an attached lever like a
bicycle handle or by a wheel as is usual in motor-
cars. Different combinations of these motions are
used by different constructors. It is of course a great
advantage if one of the pilot's hands is left free.

The steering of an aeroplane to the right or left
round a curve is a complicated operation. The
steering of a motor-car seems to be effected by
merely turning the steering wheel, but the principal

factor in turning is the sideways pressure exerted by
the road on the wheels. This becomes obvious when
a slippery road refuses to do its part and in an
attempt to round a curve the wheels sideslip. Any
body moving in a circle requires a force to act on it
towards the centre of the circle—a weight, fastened
to a string, the end of which is held in the hand, and
whirled round in a circle furnishes the simplest illustration. The curves of motor and cycle tracks and of
railways are banked in order that the pressure of the
road or rail on the wheel may supply the necessary
force without any tendency to sideslip. This suggests
the method of applying the force towards the inside
of the curve round which the aeroplane is travelling.
Just as the banking cants the motor-car the aeroplane
is canted, the wing to the inside depressed and that
to the outside raised. The air-pressure acts like
the pressure of the track and supplies the necessary
force. While the aeroplane is going round the curve
the Lift on the outer wing is greater than the Lift
on the inner because of the greater velocity of the
former, consequently the aeroplane will be canted in
the same direction. This cant, however, will be far
too great and it must be limited to the correct
amount by the use of the lateral control. As the
aeroplane moves round the curve the head must be
kept turning towards the right direction, so the
vertical rudder is also employed.

The operation of turning a corner may be divided into three parts—changing the aeroplane's path from straight to curved, going round a portion of a curve, and altering the path again to straight.   In the first and third of these the lateral and directional controls must both be used, and there is room for the display of skill in combining the two correctly.   The vertical motion is another complication to be considered. Only part of the Lift is employed in sustaining the weight of the aeroplane, consequently either the velocity must be increased or the aeroplane must descend.   Some loss of height at the beginning of the turn seems to be inevitable.   Even if the engine is accelerated, it must take some time for the aeroplane to pick up the additional speed required.

What has been said above applies to the usual form of wings.   In Lieutenant Dunne's aeroplane, as already described, the wings are twisted so that the air-pressure on the tips acts downwards.   In rounding a curve the outer wing would be depressed and the inner raised, and the lateral control must be used in the opposite direction to that described before.

The problem is suggested :—Can the steering of an aeroplane round a corner be made as simple as the steering of a steamship or a motor-car ?   It might be possible.   Perhaps two or more rudders linked up together might be used, or the wings might be twisted and curved so that the aeroplane takes the correct tilt automatically. The ideal aeroplane may be conceived as one possessing inherent or automatic longitudinal stability, in which ascending and descending could be accomplished by engine control alone, while steering to the right or left would be effected by a lever or wheel capable of one kind of motion only. If the engine were reliable, driving such an aeroplane would be almost as simple as driving a motor-car or motor-boat. Prizes might be offered for the best flights both in a straight line and round curves made under such conditions. The other controls, if any, could be sealed during the trials to make it impossible to use them without breaking the seals. Such trials would probably result in great improvements in stability and steering.

Automatic stability is a development of the method of personal control which has not yet been successfully employed. Instead of being put in action by the pilot the rudders and ailerons would be actuated by a pendulum or gyroscope either directly or through the intervention of some source of power. This system has the advantage of not requiring constant attention on the part of the pilot. Many authorities are of the opinion that the best solution of the problem of lateral stability is to be looked for in this direction.

CHAPTER IV

MODEL AEROPLANES AND GLIDERS

THE construction of a fullsized aeroplane is an expensive and laborious undertaking, but the efficiency of any new design can be tested by models or gliders. A model aeroplane is a small aeroplane which does not carry a pilot. It cannot, therefore, be controlled during flight and must possess inherent or automatic stability. The importance of this sort of stability is referred to in another chapter. By devoting attention to model flying the stability of aeroplanes could probably be greatly improved.

It seems likely that the same laws connect the dimensions, speed, and weight of models in aviation as hold in mechanics in general. A model should, strictly, have every part made to scale out of the same material as is used in the fullsized machine. The weights of corresponding parts will then be proportional to the cube of the linear dimensions. The weights per unit area of their sheets or fabrics would be proportional to the first power of the linear dimensions.

The speed will then be proportional to the square root of the linear dimensions. The weight of a quarter-scale model, for example, will be one-sixty-fourth, and its speed half that of the fullsized aeroplane.

If a model has a span of four feet the fullsized aeroplane with a span of forty feet ought to weigh 1000 times as much as the model and fly at a little more than three times the speed. A model such as is described above may be called an ideal model. In the actual model the correspondence as regards weight need not be so exact. The weight may be distributed in a different way and the internal details may be different. The total weight however of the actual model must be the same as that of the ideal model. The balance, too, must be the same both as regards the fore and aft position and the height of the point about which the model will balance. The moment of inertia should remain unaltered, the weight being so distributed that the actual and the ideal model will be turned by the same force in exactly the same way about any axis. The external surface must be unaltered that the air-pressure may correspond to that on the fullsized machine. The speed of course in the actual model must be proportional to the square root of the linear dimensions, when its behaviour as regards stability will be the same as that of the fullsized aeroplane. The ideal model, it is important to notice, is stronger and better able to withstand the

5—2

strains imposed than the fullsized machine. The
tension or compression in any part varies as the cube
of the linear dimensions, while the capacity to withstand tension or compression varies as the area of
the cross section, that is as the square of the linear
dimensions. The model thus does not test the
strength of the aeroplane or the ability of any part
to bear the strain put upon it without yielding or
breaking. This important point must be taken into
consideration in deciding the practicability of design-
ing a fullsized machine on the same lines as the
model. On the other hand, a model of a successful
fullsized aeroplane would always be able to fly if it
possessed stability and had an engine of the requisite
power.

Many competitions are now held and many prizes
offered for model aeroplanes. It would be desirable
if at these more attention could be devoted to designs
which, enlarged to scale, might be utilised for fullsized aeroplanes carrying a pilot. If some simple
rules were devised which would distinguish models
of this class from others, and separate competitions
were held for the former, more would probably
be done for the improvement of aeroplanes than is
achieved by the aviation meetings which are now
common, where the skill of the pilot and the organisa-
tion of his staff of mechanics count for so much.

A model may be flown as a glider or be driven

by a motor and propeller. Practically, twisted strands of indiarubber form the only motor in ordinary use for models. Attention may however be drawn to the successful flights of Langley's models (see Chapter VII). Langley used both steam and petrol motors and seems to have evolved suitable types of both. Possibly if motors similar to those used by Langley were now placed on the market, they would be in demand; and presumably the original motors and Langley's designs are still in existence. Langley developed his aeroplane by the method here advocated, constructing successive models of increasing size. At last he built a quarter-scale model of his fullsized machine and this model was quite successful. The fullsized aeroplane itself seems to have required only a device for launching to be a complete success, and developed as it was from a study of models, it must have had great inherent stability. Even now, if some person or institution were to put the finishing touches to Langley's inventions and give them to the world, it is quite likely that they would suffer nothing from comparison with anything done since Langley's experiments were interrupted.

The term glider when used without any qualification refers to an aeroplane carrying a pilot but without any means for propulsion. In both these respects a glider resembles a toboggan and, like a toboggan, can only progress down a slope. It differs from a toboggan

in that it makes its own path through the air and
will only come down a path of one particular inclination. If a glider at any time is moving downwards
too steeply, its speed increases as does that of a
toboggan coming down a steep hill. This will increase
the air-pressure and the glider will be lifted up, as
explained already in the case of the aeroplane. On
the other hand, if the path of the glider is not steep
enough its velocity will fall off, the air-pressure will
diminish and consequently the glider will fall. In
both cases the path automatically adjusts itself to a
particular inclination.

A glider is acted on by its weight and by the
air-pressure which, as before, may be supposed to
consist of the Drift in a direction opposite to that
in which the glider moves through the air, and the
Lift at right angles to this direction. The Lift here
is not really a Lift at all, for it does not act vertically
upwards ; but the use of the term may perhaps be
justified by the consideration that if the direction of
motion were horizontal it would act vertically upward,
and that it is most convenient to resolve the air-
pressure into components as has been done above.
These three forces, Weight, Drift, and Lift must
balance, or to state this in another way, the two forces
Weight and Air-pressure must balance (Fig. 12). The
smaller the Drift in proportion to the Lift, the more
nearly will the directions of action of the Lift and the

Weight coincide. Mechanical considerations show that the ratio of the vertical distance traversed by the glider to the horizontal distance traversed in the same time (the tangent of the inclination of the path to the horizontal) is equal to the ratio of the Drift to the Lift. In this way the ratio of Drift to Lift can be

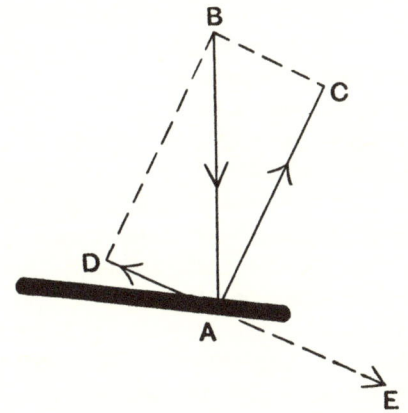

Fig. 12. *AE* direction of motion of glider, *BA* Weight,
AC Lift, *AD* Drift.

ascertained and the suitability of the glider as a model for a power-driven aeroplane tested. The more gradual the descent of the glider the less is its ratio of Drift to Lift. Unlike the aeroplane there is no limit to the ratio of Drift to Lift in the case of a glider ; but the larger this ratio is, the steeper is the descent of the glider, and the shorter the flight that can be obtained from a given height. The high degree of finish necessary to diminish the Drift of an aeroplane is not absolutely essential in a glider. Other reasons combine to make gliding easier for the amateur than experimenting with power-driven aeroplanes. The expense of an engine is saved. Less weight has to be carried, consequently the glider is both smaller and flies at a lower velocity. This reduction in speed makes the glider safer and especially lessens the risk of damage in alighting.

Gliders may be started from any high position, the top of a cliff for example, but it is safer to choose the summit of a hill whose slope is nearly the same as that of the glider's path through the air. The glider will then move parallel to the slope of the hill and never be far from the ground. It is an advantage to make the glide against the wind, for the glider and the air must always meet at the same rate and the speed of the glider over the ground is decreased. This increases the duration of the glide and makes starting and alighting easier. If the wind has an upward trend, the rate at which the glider descends is diminished and the length of the glide is increased.

Gliding is a sport within the reach of everyone. The material for a glider costs only a couple of pounds and many have been constructed by amateurs. To support the pilot at a low velocity a large amount of surface is necessary, and this suggests a biplane form of construction. Another argument in favour of the biplane has been mentioned before, when reference was made to the ease with which the necessary strength can be given to the frame. The main planes should be constructed first. Next come the contrivances, if any, for ensuring stability such as tails and fins, also the controls if these are used. When the glider is finished the balance can be adjusted by shifting the position of the pilot backwards or forwards as may be necessary. The best height for the pilot's position can also be found by trial. It may be low beneath the lower plane, level with it, or even above it. To ensure stability any of the devices described in the chapter dealing with that subject may be tried. For control any of the methods of controlling fullsized aeroplanes may be used, but simpler methods may also be adopted. It is simplest to control the glider by altering the balance. This the pilot does by moving his own body. If he moves forward he brings the head of the aeroplane down, and if backward he brings the tail down. Lateral control is effected in the same way by moving towards that side which is to be depressed. The early gliders supported themselves by their arms in an upright position and effected control by motions of their legs.

There are two well-known types of biplane gliders.

The Wright type is a biplane without any devices for securing stability, and this necessitates controls. An elevator is fixed in front, wing-warping is used for lateral control, and a vertical rudder behind for directional control. The Voisin type has a considerable amount of stability. To the main surfaces is added a large box-like tail at a considerable distance behind, and vertical surfaces between the two main planes, the whole being on the same lines as a Hargrave box-kite. An elevator in front and a vertical rudder attached to the tail may be added or the alteration of balance may be relied on for control. Skids for landing on may be attached below the main planes in any type.

During the first flights the pilot is often assisted to control his machine by men running alongside and holding ropes attached to the glider. If the wind is strong enough the glider may be anchored by a rope and supported like a kite. Practice in control could thus be gained.

The way in which the aeroplanes now in use were developed by gliding experiments is described elsewhere. A would-be aviator would still find it best to begin with gliders. On the grounds both of expense and safety it is undoubtedly the best way of learning how to use the different controls. When he has complete control over the glider the transition to the control of a power-driven aeroplane is a comparatively

small step.   Designers of a new type of aeroplane
or of new features could also test their ideas in this
way with the exception of questions connected with
propulsion.   It is to be remembered that aeroplanes
which have not inherent or automatic stability cannot
be tested by means of models as such models will
not fly.   These types of aeroplanes as also methods
of controlling and steering any aeroplane may very
conveniently be tested by gliders.

CHAPTER V

AEROPLANES

THE Lift and Drift, the stability, and the control of aeroplanes have been discussed. The gliding flight of power-driven machines, the effect of the propeller, and the problems of starting and alighting are still to be considered.

People imperfectly acquainted with the principles of flight have often suggested that parachutes should be carried on aeroplanes to provide for the safety of the pilot.

A little consideration shows that parachutes are quite unnecessary. In a properly constructed aeroplane the breakage of a part in mid-air would not occur. It should be regarded as inexcusable. The stoppage of the motor through some defect or through the exhaustion of the supply of fuel is a contingency that must be provided for. The aeroplane itself in that event forms a glider and a descent can be made along a gradually sloping path. By steering to one side or the other or by describing a spiral, a landing can be made at any chosen point within a certain range. The flight should always be so high that a suitable landing-place is within gliding distance. Besides at greater heights the air is less disturbed by the inequalities of the ground, and a good rule for pilots would be :—the more dangerous the ground surface is, fly the higher.

Variations in the working of the propeller alter the conditions of flight in more ways than one. The effect on the vertical direction of the aeroplane's path has been already discussed. Three other effects must be considered : the canting effect, the gyroscopic effect, and the tilting effect, as they may be called. The last is especially important in connection with the gliding flight of an aeroplane.

The propeller is turned round by the action of the engine ; as a result there must be a reaction of the propeller on the engine and the aeroplane to which it is attached, tending to turn the latter round in the opposite direction, that is, to cant the aeroplane. To counteract this something must tend to cant the aeroplane in the opposite direction. That side of the aeroplane on which the propeller blades move downwards may be made heavier than the opposite side, or the Lift on the former side may be made less than that on the other. When the force which the engine exerts on the propeller varies, the reaction on the aeroplane also varies and the two sides will be no longer balanced. A weight must be moved to one side or other, or the Lifts on the respective sides adjusted by the lateral control until the balance is restored.

The gyroscopic effect of the propeller is one that acts only when the propeller is tilted or turned. Any body—a top, a flywheel, or a propeller—rotating rapidly about an axis, forms a gyroscope. Certain effects are observed, if an attempt is made to alter the direction of the axle. First, it becomes more difficult to change the direction as the gyroscope rotates more rapidly. In the second place, the axle, if not prevented, in addition to moving in the desired direction, will also move in another direction at right angles. For example, the axis of a propeller in an aeroplane points fore and aft. When the aeroplane turns to the right or left, the gyroscopic effect of the propeller makes it tilt at the same time, while if the aeroplane tilts, the gyroscopic effect makes it turn to one side.

The control or steering of an aeroplane, when the propeller is revolving, requires a greater exertion on the part of controlling surfaces and is more complicated.

What has been said applies to a single propeller. Both the canting and gyroscopic effects are reversed on reversing the direction of rotation of the propeller. So far as these effects are concerned, the use of two propellers rotating in opposite directions eliminates both, and the conditions of flight remain the same no matter at what rate the propellers revolve. The Wrights, adopting this principle, have two propellers driven by chains from the same shaft, one of these chains being crossed so that the propellers rotate in opposite directions. Almost all other aeroplanes have a single screw. On an aeroplane whose motor has been fitted with revolving cylinders, the gyroscopic and canting effects would be greatly increased.

The tilting effect of the propeller depends only on the height in the aeroplane at which its axis is situated. If it is low, the thrust of the propeller tends to tilt the aeroplane head up and tail down. If it is high up, the tilting tendency is in the opposite direction. In each case the distribution of weight and of air-pressure must be such that on the whole no tilting is produced. When the propeller thrust is removed (in gliding flight), the air-pressure must be distributed in a different way to preserve the balance. The aeroplane would then meet the air in a different position, and would in general fly at a different rate, and gliding would be different from horizontal flight if the balance is such that any steady gliding flight is possible. The designer must take these two kinds of flight into consideration and make the aeroplane strong enough to bear the stresses in each.

In one case, however, things become simpler. If

the propeller axis is placed at a certain mean height neither too high nor too low, it has no tendency to tilt the aeroplane, and its working has no effect on the balance. In that case the balance and distribution of air-pressure will be practically the same in gliding as in horizontal flight, the direction of the flight alone being altered. This is a great simplification. It might, however, be an advantage, if the velocity in gliding could be reduced, to place the propeller so that gliding and horizontal flight would take place in different ways. This would be conducive to safety in landing.

An aeroplane, as has been seen, should be balanced so as to be capable of flight whether the propeller is acting or not, and the flight should be stable in both cases. It is possible for this to be so, and yet for the aeroplane to overturn when the motor stops. The position in horizontal flight may be so different from that in gliding flight that the transition from horizontal flight to gliding does not take place automatically ; in other words the range of stability in gliding may not include the position in horizontal flight. It is evidently an advantage, if the transition from power-driven to gliding flight can be made automatically and without demands on the skill of the pilot. In models the automatic transition is essential. Aeroplanes with a non-lifting tail seem to have an advantage, in this respect, over those with a lifting tail. The

latter are said to lose headway and fall to the ground
tail first when the motor stops. Of course the pilot
may bring the aeroplane into the correct gliding
position by using the elevator. It may be mentioned
that if the balance in gliding and in horizontal flight
is different, very probably the elevator will have to
be carried in a different position in gliding.

Since the support of an aeroplane depends on its
speed, it must be launched at its flying speed. A
more powerful force is required to start the aeroplane
than to keep it flying once it is started, and the more
powerful the force, the shorter the time and space
necessary for getting up sufficient speed. Langley
was unsuccessful in launching his full-sized machine
in 1903, as the catapult-like arrangement of springs
and guides which was tried did not work, although
with models it had been successful. The Wright
brothers, like Langley, at that stage in the development of the light motor, were reasonably satisfied
with a motor that developed power enough to keep
their aeroplane in flight. They also used external
aids for launching. The aeroplane was mounted on
a carriage which ran along a rail and was towed by a
rope, the power being supplied by a falling weight.
When the end of the rail was reached, the aeroplane
rose into the air leaving the carriage behind. Such
an aeroplane can only start from a place where
launching apparatus has been provided, and it is

helpless if it lands at a distance from a launching station. Later with more powerful engines Wright aeroplanes first of all ran down the starting rail driven by their own propellers, and afterwards were fitted with small wheels to enable them to start by running along the ground.

The aeroplanes developed in France, in 1906 and succeeding years, were all mounted on wheels and got up speed by running over the ground driven by their own power. This requires a more powerful motor, but the aeroplane can start from any open smooth ground of sufficient extent. This method is now universally adopted. It has the further advantage that a pilot can learn how to control the aeroplane while running over the ground, next taking short jumps and gradually increasing their length. In mounting the wheels care must be taken to guard against their buckling under the action of a sidewise strain.

Landing is a much more difficult problem than launching. Langley postponed the solution, making all his flights over water. The Wrights mounted their aeroplane on a pair of long skids or skates curving up in front as far as the elevator. For landing this is undoubtedly the best method. The skids can withstand a considerable shock, they slide only a short distance before coming to rest, and they bridge over inequalities of the ground so that a landing can be made in rough places. The French aeroplanes referred to had to land on the wheels they employed for starting. Springs or shock-absorbers are used in this system to take up the shock gradually. Even so, wheels cannot withstand so great a shock as skids; inequalities on the ground are very dangerous to them, and they do not bring the aeroplane to rest as quickly as skids do. Henri Farman was the first to combine wheels for starting with skids for landing, and this combination has become very popular. The wheels are attached to the skids by springs, so that the shock of landing pushes the wheels above the skids. Even so, after the first shock is over, the aeroplane must run on the wheels. In addition, the latter are exposed to the danger of breakage, and for this reason the combination for landing purposes is inferior to skids alone. At least one aeroplane has been fitted with disappearing wheels, that is, the wheels are drawn down below the skids for starting, and afterwards pulled up again by releasing a spring.

Accidents to aeroplanes are most frequently caused by contact with the ground. From the frequent mention in the press of breakages it would seem that it is rather the rule than the exception for some part or other of an aeroplane to give way in landing. Such accidents show a deficiency either in the construction of the aeroplane or in the skill of the pilot.

6—2

If the skids or wheels which first touch the ground are too far back, the aeroplane may fall forward—a rather common occurrence. Skids should project in front as in the Wright and Sommer aeroplanes. It is evidently very important, if aeroplanes are to come into general use, that under ordinary circumstances and in the hands of a competent pilot landing should be attended with no risk of breakage. In competitions all of which are, nominally at any rate, supposed to have the development of the aeroplane as their chief object, breakages in landing ought to be severely penalised, if they do not disqualify a competitor altogether. It would seem that in competitions such breakages, if not too serious, are no drawback to a competitor who has sufficient capital at his disposal to keep a large stock of spare parts and a staff of mechanics on the spot to make immediate repairs.

An aeroplane comes to the ground in a sloping path with both a horizontal and vertical velocity. The vertical velocity is destroyed by the shock of contact with the ground and, with skids, a part of the horizontal velocity also. The aeroplane finally comes to rest after rolling or sliding some distance. The method of alighting most in favour seems to be to approach the ground with the engine working, skimming over it as nearly horizontally as possible. At the end of a gliding descent the engine is usually started again for landing. By the use of the elevator without the engine, the path of the aeroplane could be brought nearer the horizontal, but headway would be lost at the same time and the accompanying loss of Lift would make the aeroplane fall. The motion is in fact changed into a wave-like one or a sort of jump. Unless the elevator is used at exactly the proper time, the result may be an increase of vertical velocity and a greater shock on reaching the ground. It requires greater skill to utilise a wave-like jumping motion of this sort than the continuous motion of the method first described. The gliding descent, particularly with high speed aeroplanes, is too abrupt to land with. On rough ground or in a confined space, especially when landing on wheels, it may be desirable to make the horizontal velocity as small as possible. Skilful pilots by using the elevator can bring an aeroplane practically to a standstill within a few feet of the ground, finishing with a short vertical descent.

From its very nature the aeroplane has certain limitations. It cannot hover over a spot as a dirigible balloon can, in fact its speed through the air can only be altered very slightly. It cannot rise or descend vertically but only in a sloping path. It cannot start without first getting up its flying speed, and can neither start nor land in a too confined space. Its speed makes the operation of landing one of some difficulty and of danger under unfavourable circumstances.

With one exception present-day aeroplanes are either monoplanes or biplanes. The difference is not one of principle but solely of convenience, yet in appearance there is a great difference between the two types.

Monoplanes usually have a long narrow boat-shaped framework of wooden latticework and wire, wholly or partly covered with canvas. This is usually referred to by its French name 'fusellage.' This carries the engine in front and the tractor screw is fastened to its bow. At the rear are attached the elevator, rudder, and stabilizing surfaces. The wings are attached to the fusellage near the front, and underneath are the wheels or skids or combination of both for landing and starting. The pilot and passengers have seats inside the fusellage.

The Antoinette monoplane is of this type.

In the photograph (Fig. 13) the boat-shaped body, the wings, the stabilizing tail, and the wheels can be clearly seen. The blur in front is made by the rapidly revolving tractor screw. The dihedral angle formed by the wings in the Antoinette is also seen. The great steadiness and excellent performances of the Antoinette in gusty winds have been referred to already.

The Hanriot monoplane is of the same general type as the Antoinette. The body is constructed entirely of wood in the same way as a racing skiff.

Fig. 13. Latham in flight (Antoinette monoplane).

The undercarriage is a combination of wheels and
skids. In steadiness the Hanriot is said to surpass
even the Antoinette. The Tellier is another monoplane of the Antoinette type.

The Bleriot monoplane differs in many respects
from the Antoinette. The fusellage has a square
instead of a triangular section, the horizontal tail is
lifting instead of non-lifting and the elevator is placed
on each side of the tail instead of behind it. No
vertical fin is carried in front of the rudder. The
wings are not set at a dihedral angle. In details the
two aeroplanes are very different. A Bleriot two-
seater is also constructed with a non-lifting tail. At
the Rheims meeting in 1910 the Bleriot monoplanes
fitted with Gnome motors carried off the prizes both for
duration and speed. For speed it is still unsurpassed.

All these monoplanes are controlled by wing-
working and by the use of the elevator and rudder.

In some other monoplanes as Santos Dumont's
Demoiselle, or the Grade monoplane, the fore and aft
body is not so conspicuous. These seem to consist
of a pair of wings mounted on a landing carriage,
carrying an engine, propeller and pilot, with an outrigger to carry the tail. The Valkyrie monoplane,
constructed in England, has no tail. The skids are
long and turned up in front to carry a fixed plane
tilted up as explained in the chapter on stability.
Twin propellers are used instead of a single one.

Fig. 14. The late Mr C. S. Rolls in flight on the Wright biplane.

Biplanes present an appearance totally different from that of monoplanes. The latter are more or less birdlike and have an air of simplicity which is more apparent than real.   It requires a great amount of ingenuity to make wings, which project far out, and are so stiff that they will not be bent by the air-pressure.

In biplanes the main portion appears to be formed by the upper and lower supporting surfaces. These are connected by vertical struts and diagonal tie-wires so as to form a very rigid girder.   This method of construction was first introduced by Chanute in America.

The construction of a biplane is a much simpler matter than the construction of a monoplane.   Engine and aviator are carried between the two main planes, immediately behind them is the propeller, and below them is the undercarriage for starting and alighting. Outriggers carry the tail and elevator.

The photographs of the Wright and Farman biplanes illustrate these points very well with the exception of the diagonal tie-wires.

The photograph of the Wright aeroplane (Fig. 14) shows the original type. The skids, the biplane elevator, the small semicircular vertical fins in front, and the two rudders behind, can be clearly seen in addition to the position of the pilot and engine.

The position of one of the two propellers can be made out by its axle with the toothed wheel where

it is driven by a chain.   The other propeller occupies
a similar position on the other side.   Modifica-
tions have recently been made : wheels have been
added below the skids, and an additional horizontal
surface fixed behind the rudders, acting at first in
conjunction with the front elevators and afterwards
superseding these altogether.   The skids still carry
small vertical fins in front.   The length and strength of
the skids are noteworthy features in this aeroplane,
and the manner in which they are made to form the
front outrigger is seen in the photograph.   This
biplane is controlled by wing-working and by the
use of the elevator and rudder.

The other biplane of which a photograph is shown is
that constructed by Henri Farman (see Frontispiece).
The elevator is in front, the biplane lifting tail is comparatively far behind.   The twin rudders are between
the planes of the tail.   The single propeller is seen in
rapid motion, mounted level with the lower plane.
The landing carriage is formed by two skids with
a pair of wheels attached to each by indiarubber.
Lateral control is effected by ailerons forming part
of the trailing edge of the main planes near each
extremity.   They can be distinguished in the illustration by the small gap which separates each from the
fixed part of the trailing edge, and the attached
levers can be seen by which the lower plane is moved.
In other Henri Farman types the trailing edge of

one of the tail planes is hinged and linked up with the elevator. Sometimes the span of the lower plane is shortened and sometimes that of the upper plane is increased. In a recent type the two halves of the lower main plane are set at a dihedral angle and an enclosed body provided for the pilot's seat, etc. A monoplane tail has been tried and alterations in the rudder have been made.

The Henri Farman biplane is one of a group which have been developed from the original Voisin biplane. This differed from the Henri Farman in many respects. Large vertical panels were fixed between the main planes and the planes of the tail, while ailerons were absent. The undercarriage was formed of wheels alone. The pilot and engine were placed in an enclosed body resting on the lower main plane and projecting in front to carry the elevator. In the latest type of Voisin the front elevator, the lower tail-plane, and the vertical panels are abolished. The elevator now forms an extension to the rear of the lifting tail and ailerons resembling those in the Henri Farman type are fitted.

The Maurice Farman and Sommer are other well-known biplanes of the same general type as the Voisin and Henri Farman. In England the British and Colonial Aeroplane Company of Bristol and Shortt Bros. of Eastchurch have constructed biplanes of this type.

Substituting for the lifting tail a non-lifting tail placed nearer the main planes, we get what may be called the Chanute type of biplane from its resemblance to the biplane glider developed by Chanute. The Curtis, which had such a success at Rheims in 1909, is of this type and so is the Howard Wright, constructed in England, by which the de Forest prize has been won. As has been said before, there seem to be reasons for preferring a non-lifting to a lifting tail. The latest type of Wright aeroplane with the elevator behind approaches the Chanute type.

CHAPTER VI

THE DIRIGIBLE

WE have spoken of the general principles which govern the rise of a balloon in the air, and now proceed to consider, a little more closely, the conditions that have to be fulfilled in the construction of a successful dirigible. The effect of air-resistance on a body moving through that fluid has already been discussed, and, remembering how the magnitude of that resistance depends on the shape of the moving body, the evaluation of the correct shape for the gasbag and other parts of the balloon becomes a matter of primary importance. The spherical shape is of course out of the question, and the shape which, other things being equal, seems to offer a minimun resistance, is that which approximates to the outline of a fish; not only does the pointed prow cleave asunder the air in a manner which offers a minimum of resistance, but the shape of the tail portion is such as to allow the air particles to re-unite along a line of least resistance—in fact, the shape should be so chosen that distortion of the natural lines of flow of the air should be avoided as much as possible, otherwise eddies are formed in the air, and a consequent loss ensues, energy being wasted in forming eddies which might otherwise have been used for propulsive purposes.   In an ideal shape, therefore, the entry will be somewhat blunt, whilst the tail portion tapers off gradually, in accordance with the shape of such a

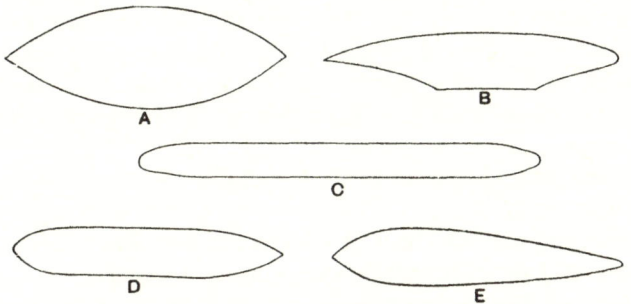

Fig. 15.   *A*. Giffard.   *B*. Lebaudy.   *C*. Zeppelin.
D. Parseval.   *E*. Renard and Krebs.

typical fish as a salmon, and with the dictum of Mr Froude that 'blunt tails rather than blunt noses cause eddies.'

These considerations as to shape apply, of course, to the car and other parts of the airship, all of which should, other circumstances permitting, be adapted to the 'streamline' form. The diagram appended (Fig. 15) shows the outline of a few of the better-known dirigibles.

Having arrived at a correct section it must be further remembered that the shape must be maintained, as nearly as possible, during the flight of the balloon. In the ordinary aerostat, ascent and descent are effected by throwing out ballast, and by allowing gas to escape ; loss of gas, however, means change in form of the sustaining envelope, and must be compensated for. In the Zeppelin airship, maintenance of form is effected by means of a rigid external framework divided by partitions into separate cells, each cell containing a gas-bag filled with hydrogen. The French school favour a non-rigid type of balloon in which the shape is kept invariable by the interior pressure of the gas, the pressure being kept up by pumping air into the gas bag to compensate for any loss of gas. The air is not pumped directly into the body of the balloon, as it would form an explosive mixture with the hydrogen ; small bags, or ballonnets, are therefore introduced into the main gas-bag, and, when any loss of hydrogen occurs, it can be immediately compensated by pumping air into the ballonnets by means of fans worked by a motor. Further, the use of several small ballonnets instead of one large one, prevents dangerous back-and-forward surgings of the enclosed air whenever the balloon is inclined to the horizontal.

The Parseval-Sigsfeld system which has been used in connection with kite-balloons provides a means of maintaining the 'shape-stability' automatically.

In this case the gas-bag of the balloon contains a single air-ballonnet having an opening O which faces the wind. The rudder bag is also filled with air, having

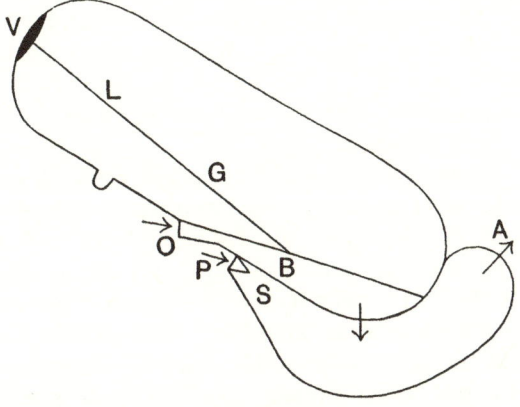

Fig. 16.
(From Moedebeck's *Pocket-book of Aeronautics*.)

an opening P near that of the ballonnet, and the excess of air flows out through an aperture at the other end of the bag. The body of the balloon is furnished with a valve V, from which runs a line LG, attached to the fabric of the ballonnet. As the balloon rises the expanding gas presses on the surface of the ballonnet and forces out any excess of air through the opening *A*. Further, the valve line is kept taut, and should the pressure get dangerously high, the fabric of the ballonnet is forced down and the tension in the valve line becomes great enough to open the valve *V*. Thus all danger of bursting the balloon fabric is automatically eliminated.

It is to be remembered that, in the application of this method to ordinary dirigibles, although it may be useful so long as the balloon is moving forwards, difficulties arise when the balloon becomes stationary, or moves backwards.

As regards motor and propeller, nothing need be added to what is said on this subject in Chapter II— here, as in the case of the aeroplane, the motor must be as light and as powerful as mechanical construction can make it—the weight per horse-power must be as small as possible. One detail we may mention in passing—the difference between the terms 'weight per horse-power' and 'weight per horse-power-hour,' which must be clearly discriminated. Thus, if a motor weighs 2000 lbs. and can develop 100 horse-power, its weight per horse-power is 20 lbs. But, in making a voyage, we have to take into account not only the weight of the motor, but also that of the fuel and other accessories of the engine. Suppose, then, to quote an example given by M. Berget, that in travelling for ten hours 2000 lbs.

of fuel are consumed. The total weight carried is 4000 lbs., in return for which the total energy developed is 1000 horse-power-hours—100 horse-power for 10 hours. Thus the weight per horse-power-hour is 4 lbs., the weight per horse-power being 20 lbs. The difference between the two phrases will therefore be evident on a little consideration. With a low 'weight per horse-power' less energy is expended in lifting the motor, and less gas is required. With a low 'weight per horse-power-hour' the rate of

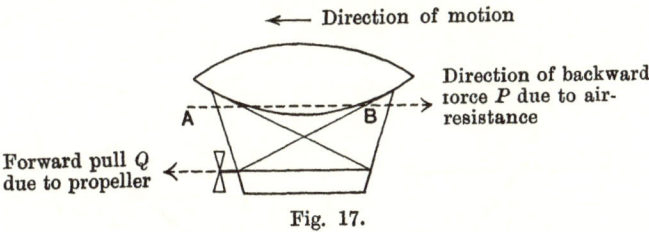

Fig. 17.

consumption of the fuel carried is correspondingly low, and the airship is enabled to make correspondingly longer voyages—that is, its radius of action is increased.

The position of the shaft of the screw is another point of importance. As the dirigible moves through the air, retarding forces act upon the balloon, the car, and its accessories, and these forces will combine to give a resultant backward force whose line of action

7—2

will be somewhere between the axis of the balloon
and the car, but nearer to the former than the
latter (AB in Fig. 17). The ideal place, therefore,
for the shaft is along the line AB, since, if the shaft
be placed anywhere else, the system of forces P and
Q will give a couple which will tend to turn the
dirigible about a horizontal axis, thus tilting the
prow of the vessel upwards. As, however, mechanical

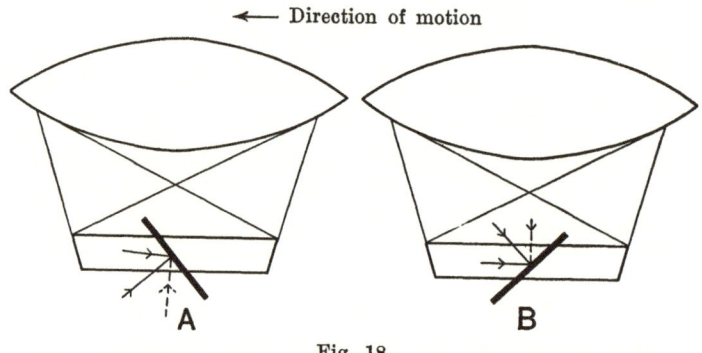

Fig. 18.

exigencies render it necessary to affix the shaft to the
car, the tilting effect must be compensated by the
use of the elevating rudder.

Having introduced the term, we may now consider
the use of this important adjunct to the dirigible.
Elevating rudders consist of pairs of planes placed on
each side of the car of the dirigible, by means of

which a fraction of the power used to drive the airship may be utilised to produce motion in a vertical direction—either upwards or downwards. As we know, when an inclined plane is moving relatively to the air, the air exerts force upon the plane, which is perpendicular to the plane. The elevating planes may be placed fore, or aft, or amidships. In the figure they are shown in the latter position; and it is clear that, with the planes disposed as shown (in Fig. 18), the air-pressure upon them will have a vertical component which in figure A will tend to raise the airship as a whole, in B, to depress it. Of course, if the elevators are placed at the stem or stern their effect is to tilt the prow of the dirigible either upwards or downwards rather than to raise or depress it bodily.

Now let us turn our attention to another point of primary importance—the question of the stability of a dirigible. Vital as the question is, an exact discussion would be accompanied by so many mathematical difficulties, that we cannot here do more than give an outline of one or two of the more important points. As will readily be understood, the main question of stability resolves itself into a discussion of what would be called in nautical language 'pitching and rolling.' Both of these movements are to be avoided as far as possible, for most of the calculations as to resistances offered to the airship when in motion are made on the assumption that the axes of the balloon are horizontal—a condition which therefore should be as nearly as possible adhered to in practice. But should any pitching (*i.e.* an inclination of the longitudinal axis) occur, the airship

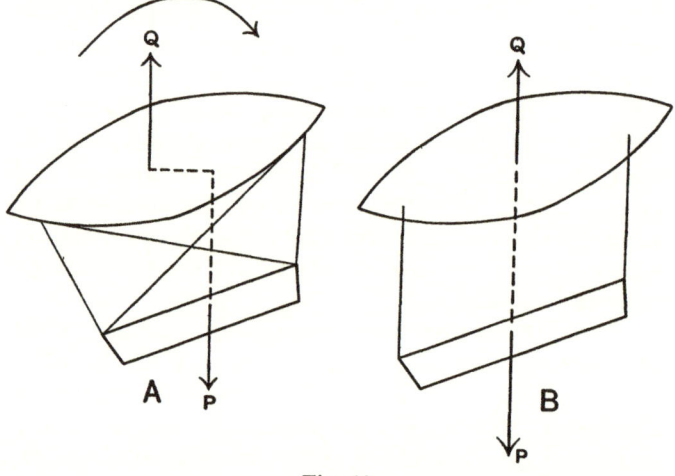

Fig. 19.

must be so constructed that the forces called into play in the inclined position will be disposed so as to bring the airship back to its horizontal position. To fulfil these requirements the centre of gravity—the point at which the weight acts—should be well below the centre of resistance, and—another point of importance—the car should be so attached to the balloon that it keeps the same position relative to the gas-bag whether the balloon be inclined or not. These conditions being fulfilled, the accompanying figure (Fig. 19, A) shows that the two forces—the downward pull due to the weight and the upward thrust due to the air displaced by the gas-bag—give a couple which tends to bring the airship back to its original position. If, however, the suspension be such as to allow the car to swing under the gas-bag, the restoring couple is not called into play, as is shown in Fig. 19, B. This preservation of relative position between gas-bag and car may be attained in two ways—either by attaching the car rigidly to the balloon, a method adopted by Count Zeppelin, or by using the so-called 'triangular' suspension, a means of preserving a rigid connection between two bodies which will readily be appreciated by any one who has done sufficient amateur carpentry to knock together a wooden framework, and put in strengthening struts and ties. A simple triangular suspension is shown in Fig. 18.

A propos of this question of stability, an important piece of knowledge is that of the time of oscillation of an airship; for as we have seen, if, in a properly constructed airship the longitudinal axis, say, becomes slightly inclined, the forces acting tend to bring the

airship back to its old position. When, however, it
reaches the horizontal position it will not stop there,
but will overshoot the mark and swing to the opposite
side, then return again, and so on. In fact, the airship swings or oscillates in much the same way, and
for the same reasons, as an ordinary pendulum swings.
Given the proper data, the period of both pitching
and rolling oscillations can be calculated. What is
the importance of this knowledge of the period?
If from some external cause vibrations or oscillations of the airship are set up, then, should the
external disturbances themselves be periodic—say,
gusts of wind arriving at definite intervals—the airship will take up the vibrations; and, if the period
of the external disturbances be equal to the natural
period of oscillation of the airship, a little consideration will convince the reader that the arrival of the
external impulses will be so timed as to continually
increase the amplitude of the vibrations of the airship until oscillations of a dangerous magnitude are
set up. This—an example of what are called
'resonant vibrations'—is a phenomenon well known
in many branches of science. A thin champagne glass,
for example, gives out a definite musical note when
smartly flipped with the finger. If this note be
loudly sung to the glass, the latter immediately
records the fact by flying to fragments—the external
waves of sound impinging on the glass when the note

is sung have the same period of vibration as the glass itself has, and the impulses produced by the impinging waves are so timed that they set the glass into vibrations of continually increasing amplitude, until the material is fractured under the strain caused by the abnormally large displacements.

Should such a source of danger arise in an airship it may be provided for by altering the position of the airship with respect to the impinging waves. The oscillations then set up by the waves are about a different axis, and the period of oscillation of the airship becomes correspondingly different, no longer agreeing with that of the external sources of disturbance.

The reader may have noticed that, in photographs of contemporary balloons, the tail end of the balloon carries either several planes arranged somewhat like the feathers on an arrow, or a number of ballonnets similarly arranged. The whole system constitutes what is known as the 'feathering,' or empennage.

It was shown by Renard, and his results have been generally verified later by Lieut. Crocco, that as the velocity of an airship increases it will reach a certain 'critical' value, above which the motion becomes unstable. Without the empennage, this critical velocity is soon reached, but by affixing these planes to the tail of the dirigible, well behind the centre of gravity, a stabilizing effect is produced, which

markedly raises the value of this critical velocity ;
i.e. it is possible to move at a much greater speed
without the motion becoming unstable.

It should be mentioned that the *value* for the
critical velocity given by Lieut. Crocco differs
essentially from that given by Renard—in fact, it is
shown that Renard's value is a particular one, only
true under certain special conditions discussed fully
by Lieut. Crocco. The paper by the latter author,
a full abstract of which is given in the recently-
published Government Blue-book, contains interesting
and valuable details concerning the effect of stabilizing
planes and the relative values of rudders placed fore,
aft, or amidships, for the purpose of steering, or for
producing lifting effects. He arrives at the conclusion
that head rudders are theoretically to be preferred,
although of course, in practice, such theoretical
benefits have often to be weighed against difficulties
of construction. The whole paper is, however, too
technical to be reproduced here—the curious (and
mathematical) reader who desires fuller information
will find it in the abstract referred to above.

Having now discussed briefly a few of the more
important points in the theory of the dirigible, we
proceed to describe shortly some of the successful
modern dirigibles. Many types are at present in
vogue, but considerations of space will hardly permit
of our giving even a brief description of each of them

Fig. 20. Airship 'Zeppelin III' over Berlin.

—we shall, then, select but one or two of the more important ones, taking first the Zeppelin balloon (Fig. 20).

This is of the 'rigid' type—*i.e.* the gas-bag retains its external shape whether the balloon be inflated or not. The external shape is 'cylindrical,' with pointed ends, the stem and stern portions having the same shape. The airship is divided into separate cells by means of partitions placed perpendicular to the longer axis of the airship, these partitions taking the form of 16-sided polygons, each of which is strengthened by means of steel wires converging on to a central hub, exactly like the spokes of a bicycle wheel. These parallel 'bulkheads' are trussed together by means of an aluminium lattice-work extending from stem to stern, and the whole is covered with fabric, which under these conditions presents a uniformly resisting surface whose external shape remains practically invariable. Inside each of the separate cells is placed a gas-bag which can be inflated with hydrogen, giving the requisite ascensional power; as the gas-bags do not completely fill the cells, an air-filled space is left between the internal fabric of the gas-bags and the external fabric of the cylinder, which, as air is a bad conductor for heat, plays an important part in preventing sudden temperature changes, and consequent sudden expansions or contractions of the hydrogen in the gas-bags. With the separate cell system it should also be noted that not only are dangerous to-and-fro surgings of the gas prevented, but the deflation of one of the gas-bags does not necessarily place the airship *hors de combat*.

Underneath the framework and running along nearly its whole length is placed a keel of triangular section made of a cloth-covered lattice-work. This feature was absent in the earlier types. At two points in the length of the keel it is interrupted to allow the introduction of two boat-shaped aluminium cars, between which, and along the keel, a weight runs on rails, whose position may be altered in order to keep equilibrium. Elevating planes are placed both fore and aft, and right at the stern is an empennage or set of two stabilizing planes inclined at about $22\frac{1}{2}°$. A triple rudder for horizontal steering is fixed at the stern between the two stabilizing planes.

As is well known, the airship has been many times reconstructed, but the arrangement of the cars and propellers—the latter are four in number, three-bladed, and about 10 feet in diameter—has remained practically the same. The horse-power of the motors has, however, been continually increased. The motors in the earlier types were each of 16 horse-power, whilst in the later patterns the motors employed were each of 110 H.P.

One motor is placed in each car ; the petrol is

stored below and transmitted to the engines by pressure. A further supply of petrol can be stored in the space in the keel below the gangway connecting the two cars. The steering is all effected from the front car, three hand wheels for horizontal steering, and two for vertical steering, being employed. The independent velocity of this great dirigible, rivalling in size an Atlantic liner, is over 30 miles per hour. The total lifting power is about equal to the weight of 34,000 lbs., of which the weight of the airship accounts for about 22,000 lbs. The useful weight which can be carried, therefore—and this includes petrol, oil, water, and crew—is not more than about 12,000 lbs.

In the section dealing with the historical development of the subject an account is given of the early activities of the French pioneers. In continuation of that sketch, mention should be made of the construction of the various balloons of the Lebaudy type, commenced in 1902.

The first balloon of the Lebaudy type had a peculiar outline which is shown in Fig. 15, *B*. A short car, two screws fixed amidships, a large horizontal stabilizing plane fixed above the car, and, leading from this to the stern, a vertical frame to the end of which is attached a vertical rudder, are among its most striking features. The shortness of the car, concentrating the weight at about the centre of the balloon, gives the gas-bag a somewhat 'broken-backed' appearance, which did not, however, apparently detract from its efficiency, for the balloon made a series of excellent flights. In 1904 a dirigible of almost the same type was constructed for the French Government. This was the famous 'La Patrie'—whose short history is well known. After effecting a successful seven-hour journey from Paris to Verdun, the airship shortly afterwards broke loose in a gale of wind and drifted away and was last seen in the North Sea. A ship of similar type was shortly afterwards designed and completed—the ill-fated République. Its excellent performances and disastrous end will still be fresh in the minds of all readers.

But of all the airships of the French school, perhaps the one which may be cited as most typical is the Clément-Bayard. We proceed, then, to analyse briefly the main points of this successful dirigible. This airship, although constructed on the same lines as the Ville-de-Paris, built for M. Deutsch by M. Surcouf, embodies certain improvements on the latter. The outline of the Clément-Bayard is pisciform, the master diameter being well towards the prow, whilst at the stern is an empennage or feathering of four pear-shaped ballonnets. This feature— what we might call a 'pneumatic' feathering, as distinct from the simple stabilizing planes described in the Zeppelin dirigible—was first introduced in the

Ville-de-Paris, where, however, the stabilizing ballonnets were cylindrical in shape. Its extreme length
is 184 feet and the master diameter 35 feet, the
cubical capacity being about 124,000 cubic feet.
Almost the whole of the neutral section is occupied
by a large ballonnet divided into two compartments.
The ballonnet can be inflated with air by means of a
single tube passing down to a fan in the car, the tube
being provided with a division enabling the air to be
pumped into either ballonnet at will. The car is
practically a girder made of steel tubing 94 feet in
length, 5 feet in height and width, the central part
being covered to carry passengers and instruments.
The motor, of 105 horse-power, drives a two-bladed
propeller, 16 feet in diameter, at 380 revolutions per
minute. The propeller is fixed at the front of the
car, and so *draws* the dirigible forward, instead of
pushing it. The car and its accessories are fixed to
the gas-bag by means of cables made of steel wire,
the ends of which are fastened to a girth sewn on to
the balloon fabric ; this dispenses with a complete
network enveloping the balloon, and enables the
fabric of the gas-bag to present a smooth surface
to the air, thereby materially reducing the air-
resistance. Elevating planes are placed well to the
fore end of the car, and a vertical steering rudder is
carried at the stern. This dirigible, carrying a crew
of six persons, made a circular journey of 130 miles—

from Sartrouville to Paris and back—in 4 hours 53 minutes, and accomplished it without discharging any ballast.

Not a great deal need be said of the three British military dirigibles, officially known as Dirigibles I, II, and III. Dirigible I—the somewhat sarcastically named 'Nulli Secundus'—had a cylindrical envelope with blunted nose and tail. The fabric was made of gold beater's skin and was unfurnished with an empennage. As is well known, after a flight to London and a descent at the Crystal Palace, the airship was moored in a gale of wind, and, to avoid a more serious loss, was summarily deflated by tearing open the ripping panel. The dirigible was afterwards reconstructed and enlarged.

Dirigible II is pisciform in outline, has a length of 152 feet, and a maximum diameter of 30 feet, the master section being rather more than $\frac{1}{3}$ of the whole length from the prow. The capacity of the gas-bag is 72,000 cubic feet. She is provided with an empennage of two pisciform ballonnets, which are in communication with the interior of the gas-bag. Forward and aft are two internal air-ballonnets into which air may be driven in the usual manner by means of a fan in the car. This latter is about 90 feet long, is built of steel tubes and hickory, and carries an 80 horse-power motor, which drives a pair of two-bladed screws, placed one on each side of the

Fig. 21. Dirigible 'Gamma' in flight.

car, and very nearly amidships. Fore and aft elevating planes are carried, and the axes of the propellers, which are 9 feet in diameter, are mounted so that they may be inclined to the horizontal, thus enabling the ship to be driven, under the power of its propellers, not only in a horizontal path, but, should occasion demand it, either upwards or downwards in an inclined direction.

Dirigible III—commonly known as 'Baby'—is a smaller pisciform ship, of 40,000 cubic feet capacity, driven by a 30 horse-power motor.

As the writer in the Aero Manual remarks, the method of work on British dirigibles, though quiet, and not at all sensational in character, is one which must be looked upon as consisting of a series of progressive experiments, each one of which gives valuable experience, and, in the end, should be capable of producing excellent results.

It will be remembered that some time ago the *Morning Post*, with laudable enterprise, organised a National Fund by means of which one of the latest and most up to date types of dirigible has been secured. This—the 'Morning Post' airship, as it has been called—is of the Lebaudy type, and, whilst it possesses the characteristics which we have already briefly enumerated, the airship embodies certain improvements, an account of which may not be without interest. The capacity of the balloon is over

350,000 cubic feet, its length is 338 feet, and its master section 39 feet in diameter. The envelope is composed of four layers of fabric—alternate layers of cotton and vulcanised india rubber—and is so impermeable that only ·6 per cent. of hydrogen escapes in 24 hours. Three air-ballonnets are fitted into the interior of the gas-bag, and, should the ballonnets be filled before the end of a flight, owing to exceptional circumstances, an arrangement is provided whereby air can be pumped directly into the main gas-bag, thus preserving the essential maintenance of external shape. The empennage consists of tail fins—not ballonnets—disposed in a cruciform shape, each fin being shaped somewhat like the wing of a butterfly. Horizontal rudders can be carried either at the bow, stern, or amidships.

The striking feature of the Lebaudy type—the long canvas-covered frame—part horizontal and part vertical, which runs for almost the whole length of the balloon immediately underneath the gas-bag, and which plays an important part as a stabilisator—is here reproduced with an important modification. In the earlier type the frame was attached directly to the gas-bag, thus impeding the free circulation of air round the envelope and tending to tear apart envelope and stabilizing plane. In the 'Morning Post' dirigible the horizontal plane is suspended from the envelope by vertical steel stays, leaving a free air

space. The stern portion of these stays, being canvas-
covered, gives another vertical plane which assists in
producing steadiness and in damping out any vibration. This alteration has enabled the breadth of the
horizontal frame to be reduced from 6 to 3 metres
without producing any decrease in the efficiency.

The two propellers—each 16 ft. in diameter—are
placed symmetrically on each side of the short car,
and are driven by two petrol motors, each of 135
horse-power, at a speed of 360 revolutions per minute.
The car itself, which is fitted up with great complete-
ness, is provided with but one landing 'pivot,' as it
may be called—an arrangement of steel tubes,
pyramidal in form, projecting downwards from the
underside of the car, and fixed well forwards towards
the bow.   When the airship descends, this point comes
in contact with the ground, forming a pivot on which
the airship can turn, so that, when anchored, it can
always be readily kept pointing towards the wind.

It is usual, in works of this kind, to attempt to
make some forecast of the uses and the future line of
development of the airship—a method which it is not
proposed to follow here, for prophecy, at most times
suicidal, is never more so than when dealing with
such a speculative subject as the future of aeronautics.
One point may be noted, in conclusion, that whatever
be the uses, military, as passenger- or mail-carriers,
or as instruments for surveying or exploration, to

which the airships of the future may be put, their usefulness, *ceteris paribus*, depends on the relation between the independent velocity of the airship—its velocity when travelling in a dead calm—and that of the wind. If the independent velocity of the airship be less than that of the wind, its range of action is restricted.

Only a series of points which lie within a certain definite angle, an angle whose magnitude can be calculated from simple mechanical considerations, can be reached by the airship. On the other hand, if its velocity is greater than that of the wind the airship can travel in any desired direction.

But, granting that the independent velocity of an airship is such as to allow of its journeying in whatever direction is desired, there is one branch of work, already an exact science, in which an airship would undoubtedly prove of great value—that of photographic surveying. By taking two photographs of a country from the two ends of a base-line of known length, an application of the laws of optics and of perspective enables us to reproduce from the photographs an ordinary topographical map of the country surveyed. And this process can be carried out very successfully by means of cameras carried on a dirigible, thereby enabling us to map out what might otherwise be tracts of quite inaccessible country.

CHAPTER VII

HISTORICAL

THE desire of flight is as old as history itself; the folk-lore of almost every nation testifies to its existence. The wings of Daedalus, with their waxen fastenings, have become proverbial. Greek myth, too, gives us the legend of Phrixus and Helle, who, borne through the air on the ram with the golden fleece, thus escaped from the wrath of their stepmother Ino. Our own Teutonic folk-lore tells how Wieland the Smith— a name which still survives as Wayland in some parts of England—after the tendons of his feet had been severed by King Nidung of Jutland, built for himself a flying cloak of feathers, by means of which he escaped to his native land. And, passing to semi-historical times, we find chronicled here and there stories, which, however ill-attested, yet testify to the strong desire inherent in mankind to conquer the regions of the air.

It is in the works of that versatile genius, Leonardo da Vinci (1452—1519), that we find the first suggestions for artificial flight that can truly be said to be of a technical character. His designs show bat-like wings, each wing consisting of several linked portions, which were attached to the body of the flying man, and were worked by the arms and legs. On the upstroke the wings closed, and on the downstroke were widely spread out. The designs never passed beyond the paper stage, and the first person of whom we have any accurate knowledge as an actual experimenter is Fauste Veranzio (1617), who made a descent from a tower in Venice in a rough kind of parachute, made of canvas fitted on to a square frame.

Besnier, a French locksmith (1678), fixed on his shoulders two parallel rods, which carried at their extremities oblong collapsible surfaces designed to act as aeroplanes. The rods could be pulled upwards and downwards by the arms and legs, so that at every upward stroke the surfaces closed much as a book closes, whilst on the downstroke they opened widely. The apparatus was afterwards modified by the Marquis de Bacqueville, who in 1742 made a more or less successful flight from the windows of his mansion, crossing the gardens of the Tuileries, and finally alighting in the Seine.

It was only natural that early attempts at flight should be made on the model of the flight of birds, but even in the early history of the science we have indications of the knowledge of the fact that the principle of Archimedes might be utilised to raise a machine in the air.

The Jesuit Francisco de Lana (1670) promulgated

Fig. 22. Airship designed by Francisco de Lana (1670).

a scheme for an aerial yacht, which, whilst utterly impracticable, is of interest as showing sound views of the fundamental mechanical principles involved.

His idea was to attach four large hollow copper spheres to a car or basket, which might be fitted with sails. The spheres being exhausted of air would give the requisite ascensional effort. He actually examined the matter quantitatively, and proposed that the globes should be 25 feet in diameter and $\frac{1}{225}$ inch in thickness. Four such globes when exhausted would be capable of raising a weight of 1200 lbs. It is hardly necessary to say that such spheres would collapse instantly under the pressure of the external atmosphere, and de Lana attempted to meet the objection by pointing out that if the spheres were perfectly uniform in quality and spherical in shape, the external pressure would merely serve to consolidate them without altering their shape—an ideal condition not to be met with outside the pages of the conventional mathematical text-book.

Pages might be filled with the fantastic visions of dreamers of the eighteenth century, but it will profit us more to pass at once to the discovery of the hot-air balloon and the hydrogen balloon in the year 1783. Stephen and Joseph Montgolfier, sons of a paper manufacturer of Annonay, a town near Lyons, had observed carefully the suspension of clouds in the air, and made conjectures as to the possibility of filling a bag with some kind of vapour which might thus be suspended in the air, much as the clouds are. They first experimented with steam with doubtful success, but finally found that a bag held over a fire, and filled with the smoke and heated vapours ascending therefrom, actually rose to some height in the air. Their experiments were immediately repeated on a much larger scale, and on June 5th, 1783, a large linen sphere 105 feet in circumference, after being filled with the heated air rising from a fire fed with bundles of straw, rose to a great height in the air, and finally fell at a distance of $1\frac{1}{2}$ miles from its starting-point, after remaining about ten minutes in the air. The news of the success of this experiment rapidly spread, and it was soon repeated in Paris under different conditions. The extremely light gas hydrogen had been discovered in 1776, and about that date Dr Black had discussed the feasibility of filling properly constructed envelopes with hydrogen and so making ascents. When the news of the success of the brothers Montgolfier reached Paris, the famous physicist, M. Charles, at once assigned the correct reason for the ascent of the hot-air balloon, that is, the relative lightness of heated air as compared with cold air, and proposed the construction of a balloon filled with hydrogen. The balloon was constructed of varnished silk, and was about 13 feet in diameter. After being inflated with hydrogen prepared by the action of dilute sulphuric acid on iron filings, it was freed, rapidly rose to a height of 3000 feet, and finally fell, after a voyage of about three-quarters of an hour, in a field distant

15 miles from its starting-point, where it was attacked by the terrified peasantry, who, under the impression that some agent of the devil was paying them a visit, dragged the balloon about at the tail of a horse until it was torn to pieces.

A few months later, Joseph Montgolfier constructed a fire balloon in Paris which made a successful ascent in the presence of the court and a large crowd of spectators, whilst on November 21st, 1783, an ascent was made in a Montgolfière by M. Pilâtre de Rozier, the first human being to ascend in a free balloon. This young aeronaut was killed two years afterwards by a fearful fall from a height of 3000 feet. He constructed a balloon which was intended to combine the advantages of the hydrogen and the fire balloon, two separate envelopes being constructed and joined together, the upper one filled with hydrogen and the lower one with heated air. Rozier's idea being that, by judicious feeding of the fire beneath the lower envelope, it would be possible to ascend or descend without losing gas. The dangers of such a construction, in face of the notorious inflammability of hydrogen, are apparent. The aerostat, after journeying safely for half an hour, was seen to be in flames and the aeronaut lost his life.

It is readily seen that the hydrogen balloon possesses many advantages over the Montgolfière or fire balloon, and the latter was soon ousted from the field. We need not however stay to consider the improvements made during the nineteenth century in ordinary ballooning,—or aerostation, as it is more properly designated,—for the fundamental principles were well established within a few years of the ascent of the first balloon, and the improvements that have been suggested and carried out are mainly in matters of detail.

The first balloon had hardly risen in the air before proposals for constructing balloons that should be dirigible were set afoot. These early ideas were, as was naturally to be expected, extremely crude; indeed, it was for long thought that an ordinary spherical balloon might be rendered navigable by fitting it with oars, sails, and the like. But as we have already seen, it is impossible in the first place to make a balloon dirigible until we have some means of communicating to it an independent velocity, and in the second place, the enormous resistance which the air exerts on a balloon of spherical shape makes it quite unfit for use as a dirigible balloon. So far as we are aware, the earliest memoir which presents clear ideas as to what is necessary was read before the Academy of Science in Paris by Brinon in 1784. He advocated the use of a cylindrical balloon with conical ends, and proposed to employ oars as a means of propulsion.

But amongst the pioneers in this field none is more deserving of notice than General Meusnier

(1784), killed at the siege of Mayence in 1793.   It is very remarkable that most of the important points which we have already indicated in the construction and management of a dirigible balloon had been

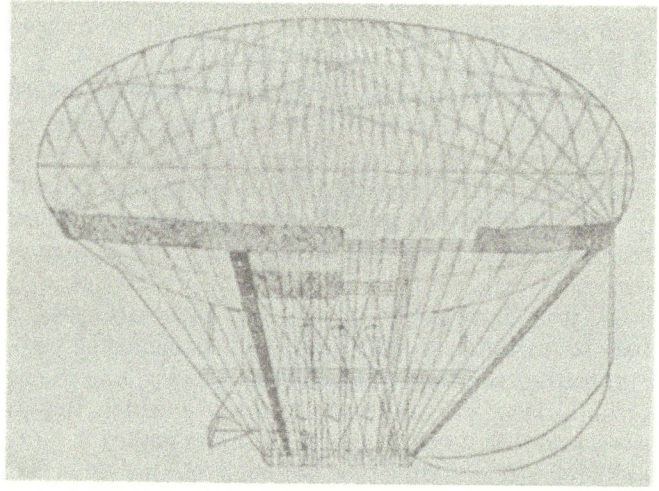

Fig. 23.   Dirigible designed by Gen. Meusnier.
(From Moedebeck's *Pocket-book of Aeronautics*.)

anticipated by Meusnier at this early stage.   To quote the words of M. Berget—'He not only re-commended the elongated form ; not only conceived the girth fastening, the triangular suspension, the

air-ballonnet, the screw propeller ; but moreover
indicated the point at which the latter should be
installed.......In this way the illustrious and accomplished officer set forth in one stroke everything
requisite for aerial navigation.' A sketch of Meusnier's proposed airship is appended. It is to be noted
that Meusnier by the use of the air-ballonnet hoped
to avoid unnecessary loss of gas or ballast, and by
raising and lowering the airship to make use of the
air-currents in different strata of the atmosphere in
order to travel at will in any given direction. The
screw propellers were, of course, worked by manual
labour, and were only intended for voyaging in a calm.

In this same year—1784—a cylindrical balloon,
with hemispherical ends, was built by the brothers
Robert, of Paris. The balloon was provided with an
air-ballonnet, at the suggestion of Meusnier, and was
directed by means of oars. The first ascent was
marred by an accident, but at the second ascent, whilst
travelling in a calm, they managed to complete an
elliptical course.

But with dirigibles as with aeroplanes, the
outstanding difficulty was the non-existence of an
engine, or source of motive power, which should
combine lightness with efficiency, and it is mainly
owing to the absence of such an engine that the
brilliant discoveries of Meusnier were allowed to sink
into oblivion. The history of science is full of such

examples. We are too apt to marvel at the remarkable coincidence which attends the genesis of a discovery at the time when it is most needed, forgetting that what usually happens is that the discovery is made again and again, and time after time forgotten because it was made before the world was ready for it. So it was with Meusnier, and a similar fate attended the speculations of Sir George Cayley on locomotion by means of heavier-than-air machines, to which we shall have occasion to refer later.

Thus, though the first half of the nineteenth century was not without its projects, we can pass them over, until we come to the year 1852 and the dirigible airship of Henry Giffard.

Giffard, well known in mechanical circles for his invention of the injector for steam boilers, had, for some years before the above date, been making attempts to construct a steam engine which should have an extremely small weight per horse-power. Having succeeded in constructing an engine of five horse-power, whose weight was about 100 pounds, he thought that such an engine might be useful for purposes of aviation, and, in 1852, he built an airship at Paris, which certainly proved successful within the limits of efficiency of his motor. The airship was spindle-shaped, 144 feet long, 40 feet in diameter at its master section, and had a capacity of about 90,000 cubic feet. Fastened to the network of the balloon

were a number of long ropes which carried at their
lower extremities a horizontal pole (over 60 feet long)
from which was suspended the car. In the car
was mounted a three-horse-power steam engine
furnished with a three-bladed propeller capable of
revolving at the rate of 110 revolutions per minute,

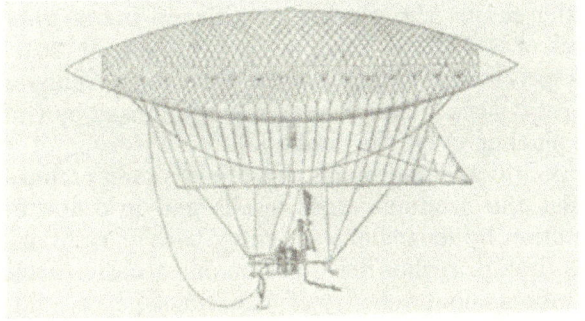

Fig. 23 *a*. Giffard's airship. (From Moedebeck's *Pocket-book
of Aeronautics*.)

whilst at the end of the horizontal rod was fixed a
vertical triangular sail to act as a rudder. The above
brief description will make it apparent that there
were, at any rate, two important defects in Giffard's
airship. In the first place, the method of suspension
employed was not one which rigidly connected the

car to the balloon, and oscillations of the car, independent of the balloon, might be easily set up. The triangular suspension, first advocated by General Meusnier, would have avoided this difficulty, though, as a set-off against this, it is to be noted that Giffard's method of suspension distributes the weight of the car and its accessories fairly evenly over the body of the balloon. Secondly, as the fate of Pilâtre de Rozier shows, the placing of a fire in the neighbourhood of a balloon filled with hydrogen or coal gas is a proceeding fraught with exceedingly dangerous consequences. The danger was minimised by Giffard by placing over the stoke-hole a covering of wire gauze which acted much as a miner's safety lamp acts, whilst the products were discharged in a downward direction by means of a chimney bent at right angles. The airship, as thus designed, on one occasion attained an independent velocity of 6 to 8 feet per second.

Three years later, new trials were made with an airship of still larger capacity (113,000 cubic feet). In order further to diminish air-resistance the shape of the balloon was altered, the length being much increased and the master-section diminished. The suspending rod was done away with, a stiff covering being carried over the upper part of the balloon and the net being attached to this. A light square car, carrying the same engine as had been previously used, was attached by its four corners to the cords of the net ; the rudder was formed of a triangular sail attached to the body of the balloon.

With this airship Giffard made a trial trip and succeeded in moving slowly against the wind. On descending, however, the airship tilted on coming into contact with the ground, placing itself with a tip upwards ; the car broke from its moorings, and the balloon was completely destroyed.

Giffard, who died in 1882, designed and constructed several other balloons, both free and captive, none of which appear to show any advances on the types which we have described.

The next dirigible worthy of consideration is that of Paul Haenlein (constructed in 1872). It presents several noteworthy points, chief of which are the curious shape of the balloon, which had the form of the figure which would be obtained by revolving the keel of a ship about an axis lying along its deck. Further, this is the first dirigible on record in which a gas engine was used as the source of motive power. The gas for consumption was taken from the balloon itself, which was inflated with coal gas. The balloon was therefore fitted with an interior air ballonnet, into which air was continually pumped in order to make good the loss of gas consumed by the engine, and to preserve the shape of the balloon. A trapezium-shaped rudder was carried aft. The balloon attained an independent velocity of 4·3 feet per second, but, although promising in many respects, lack of funds prevented the prosecution of further experiments.

As all these attempts show, it was the excessive weight per horse-power of the available motors which constituted the main check to further advances. In face of this fact it is curious to note that in 1872, Dupuy de Lôme, who had been commissioned to construct an airship by the French authorities at the time of the Franco-Prussian war, reverted to the motor, of all motors notoriously the most inefficient, as far as weight per horse-power goes—the human motor. The noteworthy points of his design are the use of the air-ballonnet, and the rigid attachment of the car to the body of the spindle-shaped balloon by means of a diagonal rope-suspension. The screws, which were worked by about eight persons, enabled the balloon to attain an independent velocity of about 8 feet a second and to deviate 10 degrees from the direction of the wind.

And now, mentioning in passing the dirigible of the brothers Gaston and Albert Tissandier, noteworthy on account of the fact that the motive power was supplied by an electric motor, the current for which was obtained from a battery of twenty-four large bichromate cells, we come to the achievements of Captains Renard and Krebs, achievements which constitute an epoch in the history of the development of the dirigible.

Their balloon (Fig. 24), which was constructed in 1884, was in shape something like the body of a fish, the master-section being at a distance from the stem about a quarter of the whole length. The car differed from those of previous types in being extremely long (about 108 feet long, 4½ feet wide and 6 feet high). A nine-horse-power electric motor, driven by the current supplied by a specially designed electric battery of chromium chloride cells, and of extreme lightness, actuated a large wooden screw propeller, which

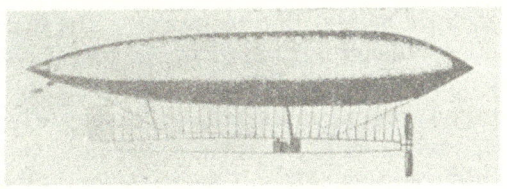

Fig. 24. Dirigible 'La France' designed by Captains Renard and Krebs. (From Moedebeck's *Pocket-book of Aeronautics*.)

revolved at the rate of fifty revolutions per minute. This propeller was fixed *forward*, so that the airship was *pulled* through the air, instead of being pushed along, as would be the case were the screw fixed aft. The rudder, which was fixed aft, was of peculiar shape, being a solid body, made of two four-sided pyramids fixed together by their bases. The car, which was fixed rigidly to the net of the balloon by means of a diagonal rope-suspension, was further provided with a sliding weight, which could be moved fore or aft to counterbalance any possible disturbances of the centre of gravity.

With this balloon, which was named 'La France,' two voyages of historic interest were made in September 1885, when the dirigible after leaving its hangar, performed several evolutions over Paris with the greatest ease, and returned to its starting-point— the first authenticated aeriel voyage on record in which a balloon has started from a definite point and returned to it under its own power. The maximum independent velocity obtained was 21 feet per second, and the machine showed good longitudinal stability.

With the account of this balloon we may fitly terminate our recital of the successes and failures of the nineteenth century, for, although several dirigibles were designed and planned during the closing years of that century, the dirigible of Renard and Krebs represents the high-water mark of engineering achievement until recent years.

The reasons for lack of further progress have been well summed-up by Major Moedebeck as follows :

(α) The laws governing the resistance of the air when acting on supporting surfaces of various sizes and shapes were not sufficiently accurately formulated.

(β) The power of the motors employed, in comparison with the great air-resistance to be overcome, was much over-estimated.

(γ) The action of the propeller was not sufficiently well-understood. And, in particular, the question as to the relative merits of small rapidly-rotating propellers, and large slowly-rotating ones, was not decided.

The brilliant successes gained in the twentieth century, and the construction and achievements of the later dirigibles have been considered in an earlier chapter. We can then, at this juncture, turn our attention to the consideration of the history of the evolution of the heavier-than-air machines—the aeroplanes and their congeners.

We have already recounted a few of the more primitive attempts in this direction, and now proceed to take up the thread of the story at the beginning of the nineteenth century. It is a very remarkable fact, that, with aeroplanes as with dirigibles, the fundamental principles and ideas were enunciated almost a century before they became accomplished facts. These principles we owe to Sir George Cayley, whose papers on the subject, published in *Nicholson's Journal* for 1809–10, still repay careful study.

After a comparison of the weight per horse-power of existing steam engines, he makes a rough calculation of the weight per horse-power of an engine which might be used as a prime mover for the purpose of aerial navigation, throwing out incidentally the suggestion that a mixture of gas and air when exploded under a piston might form a suitable source of power—a suggestion which was realised, 56 years later, in the shape of the Lenoir gas engine.

He analyses clearly the forces at work in supporting a large bird with wings outspread, gliding horizontally through the air. He performs experiments to show the variation of resistance with velocity to a surface moving through the air perpendicularly to itself. He shows that two planes inclined to each other at an obtuse angle form a laterally stable combination, pointing out, too, that the shape of the parachute which had been adopted for several descents from balloons was mechanically unsuitable. Figure 25, which is reproduced from Cayley's paper, shows this clearly. In Fig. *A*, which represents the type of parachute considered, it is clear that, the parachute being displaced from its equilibrium position during a descent, the force on the side *oa* continually increases in virtue of the fact that it becomes more and more nearly perpendicular to the current of air impinging on it, whilst the upward force on the side *ob* decreases on account of the increasing obliquity. Thus the parachute tends to tilt away from its equilibrium position. But an analysis of Fig. *B* on the same lines will show that in the displaced condition the forces called into play

tend to right the machine—i.e. the equilibrium is
stable. Further, in dealing with the stability of the
machine in the direction of its course, he showed how
the variation of position of the centre of resistance
with the obliquity of the planes could be utilized to
prevent any pitching of the machine, whilst the

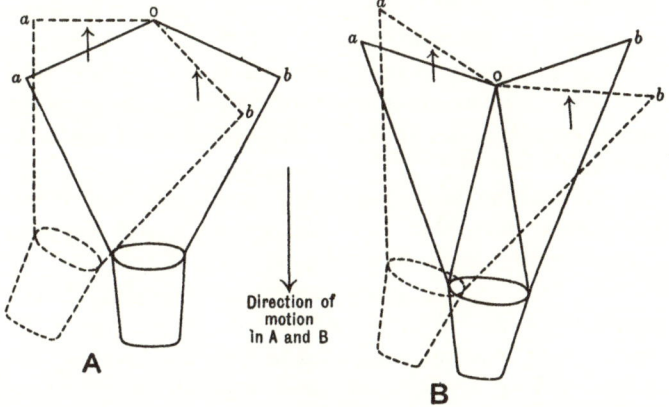

Fig. 25.

addition of a tail rudder would render the machine
perfectly stable and able to ascend and descend in its
path.

Experiments were actually carried out with a
machine having a surface of 300 square feet, and a
total weight, when loaded, of 140 lbs. It was perfectly

steady in action, and would glide down obliquely in any desired direction according to the set of its rudder.   The machine was accidentally broken before an opportunity had arisen for testing it with a propelling apparatus.

We see, then, that Sir George Cayley had very clear ideas as to the solution of the fundamental problems involved in artificial flight, and had a motor or engine of sufficiently small weight per horse-power been in existence in his day, there is little doubt that the science of aeronautics would have matured much more rapidly.   As it is, he has well earned the title of the Father of British Aeronautics.

Mentioning in passing the projected, but never completed aeroplane of Stringfellow and Henson (1843)—a structure having a superficial area of 7000 square feet, and weighing 3000 lbs.—we come to the classical work of F. H. Wenham.   In a paper on Aerial Locomotion read at the first meeting of the Aeronautical Society in 1866, after a careful and capable study of the soaring powers of various birds, he enunciates the important principle that when an inclined plane is being driven through the air the supporting force is not derived from the whole surface of the plane, but is limited to a narrow portion near the front or 'entering' edge.   Consequently supporting surfaces for aeroplanes should have their dimensions perpendicular to the line of flight much greater than their dimensions in the direction of the line of flight. Further, noting how closely a large bird in rapid flight can skim to the surface of a placid lake without ruffling it in the slightest degree, and hence proving that the stratum of air displaced by a bird in rapid flight is very thin, he pointed out that in order to obtain the necessary surface for supporting heavy weights, the long narrow planes which form the supporting surfaces may be superposed, or placed in parallel rows, one above the other, with an interval between.

These conclusions, and the equally important result that to accomplish artificial flight did not require the expenditure of such an enormous amount of power as was popularly supposed, are remarkable instances of correct mechanical conclusions being drawn from simple and direct observation of natural phenomena. Wenham's paper, which has recently been published in a more accessible form, will ever remain one of the classics of aeronautics.

The principles advocated by Wenham were carried to an extreme extent by Phillips. His flying machine was strongly reminiscent of a Venetian blind. It had a total supporting area of 136 square feet, and was always run as a 'captive,' being confined to a circular track of 100 feet radius. Whilst the machine proved its capability for rising from the ground the experiments lacked finality, for in such a machine little can be proved as regards its stability in free flight.

A series of experiments on a 'captive' machine of very large size was carried out by Sir Hiram Maxim about 1890. The machine had a total lifting surface of 6000 square feet, and when completely fitted up weighed 8000 lbs. It was driven by two specially designed engines, each weighing 310 lbs., and each capable of developing 180 horse-power. On several occasions a lifting effect of 3000 to 4000 lbs.-weight was obtained, but the experiment was finally abandoned, mainly owing to its costly character.

The next development in the history of artificial flight takes us across the Atlantic to review the work of the American physicist S. P. Langley.

The name of Langley has long been associated in the scientific world with epoch-making advances in solar physics, but it was not until the last ten years of his life that his name was associated with the subject of artificial flight.

The striking feature of Langley's work is its remarkable originality. In attacking the subject nothing was taken on trust, but the work was begun at the very beginning. He first instituted experiments to test the amount of mechanical power requisite to sustain a given weight in the air when advancing at a given speed, and having satisfied himself that mechanical flight was possible with the power at his command, he entered upon the second stage of his investigations—the acquisition of the art of directing that power.

After numberless experiments with models driven by means of twisted rubber, he turned his attention to the construction of a model which should be driven by a steam engine. Here again disheartening difficulties were encountered, chiefly in the matter of making an engine which should be efficient and at the same time of low weight per horse-power, and it was not until four aeroplanes—or, as Langley called them, aerodromes—had been constructed that these constructional difficulties were overcome. The fourth aerodrome seemed to give promise of success, and was taken 30 miles down the Potomac to a houseboat from the upper deck of which the launches were made. And now a new series of difficulties presented themselves in the matter of launching the aerodrome, and when these were finally overcome and the aerodrome was successfully launched, it behaved in a most erratic and disappointing manner, now plunging forwards and downwards into the river, now soaring upwards until its wings made so steep an angle with the horizon, that, unable to sustain itself, it slid backwards into the water.

These defects were traced to the irregular deflection of the wings caused by the pressure of the air upon them, and again an exhaustive series of experiments had to be performed in order satisfactorily to strengthen the wings without unduly over-weighting the model. At last, after three years unremitting labour in the construction of models, an aerodrome was produced which made a satisfactory flight. The story of that flight—the first of its kind—is best told in Langley's own words :—

'On the 6th of May of last year (1896) I had journeyed, perhaps for the twentieth time, to the distant river station, and recommenced the weary routine of another launch, with very moderate expectation indeed ; and when on that, to me, memorable afternoon the signal was given, and the aerodrome sprang into the air, I watched it from the shore with hardly a hope that the long series of accidents had come to a close. And yet it had, and for the first time the aerodrome swept continuously through the air like a living thing, and as second after second passed on the face of the stop-watch until a minute had gone by and still it flew on, and as I heard the cheering of the few spectators, I felt that something had been accomplished at last, for never in any part of the world or in any period had any machine of man's construction sustained itself in the air before for even half of this brief time. Still the aerodrome went on in a rising course until, at the end of a minute and a half (for which time only it was provided with fuel

and water), it had accomplished a little over half a
mile, and now it settled, rather than fell into the
river, with a gentle descent. It was immediately
taken out and flown again with equal success, nor

Fig. 26. Langley's aerodrome in flight.

was there anything to indicate that it might not
have flown indefinitely except for the limit put upon
it.'

The weight of this model was 30 pounds, and the
distance from tip to tip of its supporting surfaces was
about fourteen feet.

The remarkable success of these models led to the inception of a series of experiments for the making of a full-sized man-carrying aerodrome. It is unnecessary here to enter into the details of these experiments. Suffice it to say, that, after many vexatious delays in the construction of the engines, an aerodrome was finally completed. The efforts to launch it were, however, attended with total disaster, due to defects in the launching apparatus, and to no fault of the aerodrome; and the prohibitive cost of the experiments prevented their being carried to a satisfactory conclusion. Nevertheless, the work of Langley has been of fundamental importance in the study of aeronautics, and has paved the way for much of the later and more successful work.

With the mention of the work of one other pioneer, Ader, who in 1897 made the first known flight in Europe, we turn to consider the efforts of another class of experimenters, who attacked the problem of artificial flight in a very different manner. As we have seen, Langley, for example, experimented with motor-driven models, afterwards proceeding to full-sized machines, but other experimenters advocated preliminary experiments with gliders, arguing that small patterns or models of flying machines do not permit of extended observation, and that, for experiments in artificial flight to be instructive, a man must participate in the flight.

Following this train of reasoning, it was pointed out that attempts at flight should begin with the simplest apparatus, and that with simple wing-surfaces, similar to those of a bird, limited flights can be carried out by gliding through the air from elevated points.

Moreover, a great deal concerning the peculiarities of the wind effects, the shape of supporting surface necessary for producing the best results, and the proper conditions for stability, can be learnt from such experiments, whilst the introduction of a motor needlessly complicates matters.

The pioneer, and one of the most brilliant exponents of this school, was undoubtedly the German, Otto Lilienthal, whose experiments in this direction have become classic, and to whose work we owe much of the information of the preceding pages.

Lilienthal (1848–1896) began his work cautiously and carefully. Leaping with his glider from quite small heights (4 and 6 feet), he gradually increased the height of his 'taking-off' point, until, from a height of 30 yards or more, he found he could safely make glides of more than 300 yards, altering his direction of flight to right or left by corresponding movements of his body. He then proposed to try the effect of a motor, and constructed, for use with the motor, a double-decked aeroplane. But an unfortunate calamity brought his promising experimental work to a premature conclusion, for, while testing a steering arrangement, he was caught in an awkward current, the machine lost its equilibrium, and Lilienthal was dashed to the ground from a height of 20 yards, fracturing his spine.

His work has been carried on and extended by Herring in America, Pilcher in England, and Ferber in France.

In England, Pilcher carried out many experiments with an apparatus resembling that of Lilienthal, and, like Lilienthal, was on the point of attaching a motor to his aeroplane, when an unfortunate accident—a fall of 30 feet, occasioned by the breaking of one of the ribs of his glider—cut short his promising career. He, however, left several valuable results behind him, having demonstrated that a considerable dihedral angle in the wings produces a diminished stability in side winds, that light wheels at the front lessen shocks in landing and are convenient in the transport of the machine, and that the difficulty of controlling the aeroplane is greatly increased by unduly lowering the centre of gravity.

Of later aviators, although much original work has not been done in England, Mr Cody, Mr A. V. Roe, the only aviator who uses a triplane, and Lieut. Dunne, whose aeroplane has been previously mentioned, should be mentioned.

In America, Herring, who had been a pupil of Lilienthal, introduced his ideas to Mr O. Chanute, of

Chicago, who, in considering the difficult question of equilibrium, endeavoured to make it automatic by using movable surfaces instead of controlling the aeroplane, as Lilienthal had done, by movements of the operator. After some experiments with a Lilienthal glider, and a multiple-winged machine having eight separate planes arranged in pairs so as to form four rows, one above the other, a double-decker or biplane was constructed with which seven hundred satisfactory glides were made.

But about this time the brothers Wright were carrying out their famous experiments in gliding. Their interest in the subject, aroused by the accounts of the work of Lilienthal, led them to make a large number of gliding experiments with a biplane— experiments which elicited much valuable information. They discarded the tail, substituting for it a hinged horizontal rudder set at a negative angle of about 7°, controlling the balancing of the machine by the rudder, and steering to the right and left by warping the appropriate wing. It was further discovered that the lifting power of a full-size machine was much less than the results of Lilienthal's experiments would suggest, and that a pair of superposed surfaces has less lift in proportion to drift than either surface separately. They, too, first adopted a prone position for the operator, thus lessening materially the head resistance.

After satisfying themselves that they had complete control of the apparatus, a motor and propeller were added, and on the 17th of December 1903 a successful flight of 260 metres was made, the duration of the flight being 59 seconds.

Leaving the experiments of the brothers Wright for the moment, let us now turn our attention to the progress of the movement in France. In 1898 Captain Ferber, of the French army, endeavoured to extend Lilienthal's gliding experiments, but with indifferent success until experiments were made with biplanes resembling those of the brothers Wright. Many satisfactory glides were made, and Ferber then proceeded to fit the aeroplane with motor and propeller. The step was somewhat premature, and lack of support from the War Office brought the experiments to a temporary close. Later, in 1908, Captain Ferber rebuilt his aeroplane and made a successful flight with it. Meanwhile MM. Archdeacon and Voisin had been making successful gliding experiments with a glider resembling that of the brothers Wright, but which, as the result of experience, was altered until it resembled the form of aeroplane afterwards associated so closely with the name of Farman.

In 1906 Santos-Dumont, who has done as much as any man to popularise aviation, appeared on the scene. His aeroplane was at first attached to a dirigible balloon, but with growing confidence the balloon was discarded, and on the 23rd October 1906 Santos-Dumont won the Archdeacon prize for a flight of 25 metres—the first flight in Europe, with the exception of the half-forgotten effort of Ader in 1897. A month later, a distance of 220 metres was covered, and in January 1908 Henri Farman covered a triangular course of one kilometre, winning the Archdeacon-Deutsch prize of £2000.

And now the story of aviation in France becomes one succession of records made and broken. In July 1908 Wilbur Wright arrived, and the remarkable demonstrations given by him withdrew, for a time, public attention from the French aeronauts, until Farman established yet another record by making the first cross-country flight from Châlons to Rheims, a distance of 17 miles, and Louis Blériot made the first closed trip across country from Toury to Artenay and back, a distance of 19 miles. In July 1909, Latham attempted to cross the English Channel, and after an unsuccessful attempt was forestalled by Blériot, who made the first cross-Channel passage on July 25th. The Comte de Lambert, at the Juvisy meeting, flew over Paris and rounded the Eiffel Tower at a height of more than 1000 feet, and on the 31st of December 1909, Maurice Farman accomplished a cross-country journey of 47 miles in 50 minutes.

The achievements of 1910 are fresh in the minds of all—the flight of Paulhan from London to Manchester, and Graham-White's plucky endeavour to accomplish the same feat, the double crossing of the Channel by the late Hon. C. S. Rolls, Latham's great feat in rising to an altitude of 3,600 feet, an accomplishment since surpassed by several aviators, the record at present being in the possession of Chavez[1], who at Blackpool reached the great height of 5,405 feet—all these are matters of to-day, and, remarkable as they may be, the promise of the future is still greater.

[1] At present (April 1911) the official height record (10,746 feet) is held by Legagneux.

CHAPTER VIII

CONCLUSION

THE reader of the preceding pages may be curious to obtain some information as to the work accomplished by British men of science and mechanicians, thinking, perhaps, that England has been somewhat lax in encouraging aeronautical construction and research.

In this connection, a study of some of the more important points in the just-published report of the Advisory Committee for Aeronautics, appointed under the presidency of Lord Rayleigh on April 30th, 1909, may not be without interest.

In the first place, realising clearly that much further exact knowledge of the laws of air-resistance to bodies moving therein is highly desirable, a very complete equipment has been fitted up at the National Physical Laboratory. For experiments on the efficiency of rudders, the resistance of models, the forces acting on inclined surfaces of various forms, the distribution of pressure on inclined planes etc., a

wind channel has been constructed 4 feet square and
20 feet long. A steady flow of air for all velocities
up to 30 feet per second can be kept up in this
channel by means of a centrifugal fan, and the
behaviour of the various models, when placed in this
air current, can be experimentally investigated.

For making propeller-tests, investigating the
efficiency of various types, and examining the effect
on the efficiency of variations in blade-area, pitch,
and slip, a whirling table of special design has been
constructed, consisting of a horizontal arm 30 feet in
length, which can be rotated rapidly about a vertical
axis at speeds varying from 5 to 30 revolutions per
minute, and which carries at its extremity the propeller under test. This method has been much used
in the experimental study of the problems mentioned
above in connection with the air channel. For this
type of problem, however, the wind channel is
susceptible of higher accuracy, for, if the whirling
arm be set up inside a room, the air may be set
into circulation by the motion of the arm, and,
should the whirling arm be set up in the open air,
the results are liable to be affected by the winds
prevailing at the time of observation. But for the
study of propeller actions, in which the motion of the
air relative to the propeller is induced by the rotation
of the propeller about an axis which at the same time
advances in the direction of its own length, the

whirling table has many advantages, and it is to the study of this type of problem that it will be directed.

In order to expose large models to the action of wind, two wind towers have been erected, each about 60 feet in height, and 350 feet apart. On the top of each tower is a level platform 20 feet × 3 feet 6 inches, which can be turned about to face in any desired direction. On the ground, midway between the two towers, is an observation hut containing the recording instruments. Pressure is transmitted from the models exposed on the top of the tower by means of lead pipes running down the sides of the tower and leading by a covered trench to the hut. The velocity of the wind corresponding to this pressure is also read by means of gauges fixed in the observation hut and communicating with the tower-top by similar systems of lead tubing. In this manner simultaneous readings can be taken of the resultant pressure on any given model and the corresponding wind velocity.

The laboratory is also fitted with a complete equipment for making efficiency tests of light petrol motors, and with apparatus for carrying out experiments on the tensile strength and bursting strength of balloon fabrics.

Some experimental work of interest has been carried out on the resistance of balloon models when suspended in a current of water, a knowledge of their behaviour when placed in such a medium being of

value in enabling one to estimate their relative behaviour in air.

The model under experiment was suspended from
a weighing lever resembling a steel-yard, and placed
in a channel 12 inches in width and 7 inches deep,
through which water flowed at velocities varying from
1 to 1·9 feet per second, within which limits the
resistance was found to be proportional to the square
of the velocity. (It is to be noticed in passing that
the model is kept stationary while the water flows
past it at a known velocity. This permits of the
necessary measurements being made with greater
ease than would be the case did the model move
through stationary water.)

Various models were experimented upon, the
head-and-tail pieces generally being separate so that
any head could be screwed to any tail, and cylindrical
pieces of varying lengths could be fitted between the
head and tail when necessary.

Measuring the total resistance offered by any
model in pounds-weight, and its volume in cubic
inches, the ratio of the volume to the resistance will
clearly give a measure of the efficiency of the model,
and the greater this ratio the greater will be the
efficiency of the model. (For it is obvious that the
condition to be fulfilled is that a *large* volume should
give as *small* a resistance as possible.)

A particular pisciform shape was found to give

the best results, while the increase of resistance due
to the added middle portion was found to be directly
proportional to the increase of length.

Using the same tail with different heads, or the
same head with different tails, the effect of any given
head or tail could be estimated separately, and the
experimental results clearly show that blunted heads
or tails are eminently undesirable.

Further, the increased resistance in any model
fitted with a cylindrical middle portion being due to
surface friction, this increase can also be separated
out and estimated independently.   Let this resistance
per square foot be F pounds-weight, and the corresponding velocity of the water b feet per second, the
experimental results give

$$F = \cdot00455b^2.$$

Now, Zahm has experimented on the surface-
friction of boards placed in currents of air moving
with various velocities, and Froude has performed a
similar series of experiments in water.   A comparison
of their results goes to show that under similar
conditions the resistances in the two media are proportional to their densities.   As the density of air is
about $\frac{1}{780}$th that of water, the 'coefficient of friction'
for air should be $\frac{1}{780}$th part of $\cdot00455$, that is, about
$\cdot000006$.   For boards 2 feet long in a current of air
at 10 feet per second, Zahm found that the value of

this coefficient was about ·000007—a sufficiently close agreement.

Measurements were also made showing the increase of resistance due to the addition of ballonnets or fins at the tail end.

The water channel and balance just described have also been used to determine the relative efficiencies of various types of rudders and lifting planes for dirigibles. The rudder experimented upon was placed in the centre of the water channel and fixed to a thin steel spindle, this spindle being fastened to a vertical arm which was fixed to the knife-edge of the balance. As the model rudder was attached so that its plates were parallel to the knife-edges, the pressure of the water on the plates will tend to tilt the balance, and equilibrium can be restored by moving the sliding weight, the position of the weight giving a measure of the normal pressure on the plates. Further, by rotating the balance about a vertical axis the plate will become inclined to the current, and measurements can be made of the normal pressure on the plate for all inclinations from 0° to 90°. The results were reduced to 'air' values by means of the 'density-ratio' result which has been already mentioned.

Various forms were tested—a single plate—double plates with the planes at varying distances apart—and one or two triple plates. A comparison of the

curves drawn showing the variation of normal pressure
with angle of incidence in a current of air flowing
with a speed of 30 miles per hour shows that the
interference between the planes lowers the efficiency
of the rudder, and that, unless the two planes be
separated by a distance greater than twice their
length, a single plate is more efficient than two
parallel plates. If however the top and bottom plates
be cut away, so that the plates under experiment
are merely joined by short connecting spars the
efficiency of the rudder is greatly increased, the
increase varying from 10 per cent. at an angle of 20°
to 75 per cent. at an angle of 40°.

A model rudder of the box-kite form gave noteworthy results, the pressure intensity being greater
than that over a single flat plate up to an inclination
of 26° and greater than that over the double plate
just mentioned up to an inclination of 32°. Triple
rudders of special shape were also experimented upon
and showed peculiar fluctuations in efficiency as the
angle of incidence increased, the efficiency showing a
minimum value at an angle of 10°, then rising to
a pronounced maximum at an angle of 35°.

Of the other activities of the National Physical
Laboratory we may shortly note experiments on the
tensile and bursting strengths of balloon fabrics, on
the rate of diffusion of hydrogen gas through similar
materials, and a critical report on the mechanical and

non-corrosive properties of various light alloys which points to alloys of aluminium with copper or with copper and manganese as being most satisfactory.

A question of primary importance, and one treated at length in the report, is that dealing with the accumulation of electric charges on the surfaces of balloons, and the dangers arising therefrom. Several cases in which the balloon gases have exploded, either in the air or on landing, and which have resulted in the total destruction of the balloon fabric, are sufficient to emphasise the reality of the danger and the necessity of taking special precautions to avoid it. Different hypotheses have been suggested as to the manner in which such charges accumulate. It is certain, however, that at a height of two kilometres, electrical pressures or 'potentials' measuring as much as 100,000 volts have been registered. Now a balloon ascending into such a region would acquire the potential of that region, and if, as often happens, parts of the balloon are good conductors, and are surrounded by insulators, these conducting parts will acquire a charge which will be held—especially in dry weather—during the descent of the balloon. Thus it may happen that on reaching the ground, parts of the balloon may be at a potential of many thousands of volts, and of course on connecting these parts to earth an electric spark passes. These remarks apply specially to the dangerous region of the valve—

which, in one case quoted, for example, was of metal and one square metre in area.

The remedies proposed are various. The *whole* of the balloon should be made either conducting or insulating. If the whole of the balloon be a conductor this 'will prevent the accumulation of local charges in dangerous spots[1],' whilst, if the whole of the balloon be made non-conducting, any charge which the balloon may acquire in the atmospheric regions of high potential, will be slowly discharged in a low potential neighbourhood, without the accompaniment of sparks. 'The danger lies in having a conducting patch—the valve or a patch of damp—surrounded by insulating parts[2].'

Other observers recommend the attachment of the valve to the car—which, however, it has been pointed out, involves technical difficulties; another suggestion is that all parts of the balloon should be kept in good electrical communication, and that certain specially dangerous parts, such as the valve, should be connected to a discharger furnished with sharp metallic points, which would disperse the charge as rapidly as it accumulates, and thus prevent the accumulation of dangerous charges.

As has been said, however, the whole problem is one of great importance, and although some experi-

---

[1] *Report*, p. 121.   [2] *Ibid.* p. 120.

mental work of great interest dealing with the
question has been carried out at Glossop and
Eskdalemuir observatories, further results would be
of great value.

In the experiments mentioned it has been clearly
shown that an insulated conductor supported at a
definite height from the ground will acquire an
electric charge at a rate depending on the situation
and dimensions of the conductor. If the conductor
is then rapidly transferred to the ground without
having time to lose its charge in the descent, an
electric spark passes on discharge. Of course the
same phenomena are presented if the insulated conductor rises rapidly from the earth to a region of high
potential, as the passage of an electric spark depends
on the *difference* of potential between the body and
surrounding objects.   Although the experimental
work is more difficult, it has been similarly shown that
a non-conducting body can acquire a charge in the
same way as the insulated conductor referred to in
the preceding paragraph.

The author of this report recommends the use of a
balloon made entirely of conductors. This avoids the
danger due to sparking between different parts of the
balloon, whilst a knowledge of the rate at which the
balloon will acquire a charge, and of the rate at which
electrical potential varies as one ascends or descends
through the air, will enable calculations to be made

which will exhibit the maximum speed of ascent or descent which is allowable with safety.

It may be mentioned, in conclusion, that in addition to a number of other papers dealing with subjects of mathematical and experimental interest, the *Report* contains a large section dealing with meteorological questions, together with a series of valuable abstracts of original papers published during the last ten years.

BIBLIOGRAPHY

Chatley. The Problem of Flight. 1907.
Moedebeck. Pocket-Book of Aeronautics. (Translation) 1907.
*The Aero Manual. 1909.
*Berget. The Conquest of the Air. (Translation) 1909.
*Hildebrandt. Airships Past and Present. 1908.
Reports of the Advisory Committee for Aeronautics. (Govt. Blue Books.)
*Langley. Researches and Experiments in Aerial Navigation. Washington, 1908.
⎰Cayley. Aerial Navigation.
⎱Wenham. Aerial Locomotion.
    (Reprinted by the Aeronautical Society of Great Britain as Aeronautical Classics Nos. 1 and 2.)
Encyc. Britannica. Article 'Aeronautics.' (10th Edit.)
Petit. How to build an Aeroplane. 1910.
Lanchester. Vol. I. Aerodynamics. 1907.
            Vol. II. Aerodonetics. 1908.
Brewer. The Art of Aviation, 1910.

Journals:—
'Flight' and 'The Aero' (weekly).
'Aeronautics' (monthly).
'The Aeronautical Journal' (quarterly).

(This list merely includes a few of the many volumes dealing with the subject of Aeronautics, which may be of use to the reader who desires to read further on the subject. Those volumes marked with an asterisk are more or less popular in their mode of treatment.)

INDEX

For EU product safety concerns, contact us at Calle de José Abascal, 56–1°,
28003 Madrid, Spain or eugpsr@cambridge.org.

www.ingramcontent.com/pod-product-compliance
Ingram Content Group UK Ltd.
Pitfield, Milton Keynes, MK11 3LW, UK
UKHW010850090126
466816UK00011B/144